AN ALASKAN LIFE OF
HIGH ADVENTURE

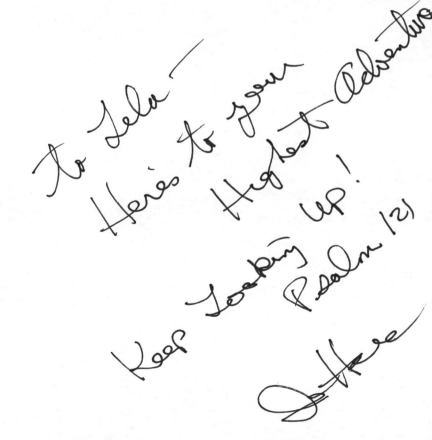

To Tela —
Here's to your
Highest Adventure!

Keep Looking Up!
Psalm 121

AN ALASKAN LIFE OF
HIGH ADVENTURE

JIM HALE

TATE PUBLISHING
AND ENTERPRISES, LLC

Published by Tate Publishing & Enterprises, LLC
127 E. Trade Center Terrace | Mustang, Oklahoma 73064 USA
1.888.361.9473 | www.tatepublishing.com

Tate Publishing is committed to excellence in the publishing industry. The company reflects the philosophy established by the founders, based on Psalm 68:11,
"The Lord gave the word and great was the company of those who published it."

Book design copyright © 2012 by Tate Publishing, LLC. All rights reserved.
Cover design by Joel Uber
Interior design by Matias Alasagas

Published in the United States of America

ISBN: 978-1-62024-403-6
1. Biography & Autobiography / Adventurers & Explorers
2. Biography & Autobiography / Personal Memoirs
12.09.14

"Coursing through the Hale family arteries is not blood but adventure—from parents to grandchildren, the arteries pump pure adventure."

Larry Kaniut, author of *Alaska Bear Tales* and *Cheating Death: Amazing Survival Stories from Alaska*

"we turned sideways to the flow...unable to breathe due to the slipstream of wind around our bodies causing a negative pressure and sucking the air out of our lungs."

"Jim Hale takes us on an enlightening thrill ride through his life: ascending high peaks, soaring with ravens, seeking good landings, and finding love and meaning through it all... Everyone is fascinated with Alaska, every Alaskan is fascinated with climbing Denali, everyone should read this spirited adventure!"

Seth Anderson,
Inventor, Co-founder of www.lokigear.com

"As an adventure educator, explorer, and history enthusiast, this is a must read. Every story ignites the innate desire to seek out adventure, to put one's foot in front of the other and pursue your passion. Hale's desire to explore and understand Alaska's rugged mountains, especially in the early years of guiding on Denali, has pushed him to extremes most could never imagine. Hale shows us that life's greatest lessons are learned through trial and error and overcoming fear. With perhaps the greatest lesson of all, the realization that Life is a gift."

Chad Thatcher, PhD
Outdoor Education Program Director
Colorado Mesa University

DEDICATION

For my dad and mom, Dr. George and Mary Hale, two diehard, never-give-up Alaskan pioneers who lived adventurously, loved Alaska and introduced and encouraged me in the pursuit of the life-changing wonders of an adventurous Alaskan life.

ACKNOWLEDGEMENTS

Life is about family, love, and passing lessons learned on to those who come along with you. It would have little meaning without someone to share it with.

To my wife Roni: It's been Jim and Roni for over thirty-five years now. I wouldn't be me without you. Thanks for making my life and this book more complete.

To my adventurous, independent, and capable kids: You inherited this adventurous spirit and carry it with you into your world. Your grandparents would be proud. So are we.

Thanks to Pastor Wayne and Marveen Coggins and Greg McElveen for your encouragement and guidance. Pat Martin and Robert Deal for your excellent copy editing. The team at Tate Publishing has been encouraging, helpful, and professional. It has been a pleasure working with you.

To all of you who have shared these adventures, thank you. Roni, kids, grandkids, family and friends, you make it all worth living and remembering.

TABLE OF CONTENTS

FOREWORD

Jim Hale and I lived within a few miles of each other, were only a few years apart, were both enthralled by Alaska, but never met until we were much older.

Jim was drawn to and became addicted to the wilderness and the adrenaline rush that comes with measuring yourself against the many facets and challenges of the Alaskan mountains. Denali is the tallest mountain in the world when measured from the actual base. It rises up from the earth and is spectacular. Everest is higher, but from base to summit much smaller.

Did you know that ravens steal food from buried, hidden caches as high as 18,000 feet—caches that may mean life or death to climbers? Did you know that flying on and around glaciers is an entirely different paradigm for bush pilots? Crashes are common due to unreadable and inexplicable wind currents and rapid temperature changes in the thin, high-altitude frigid air which is as deceptive as a card dealer in Vegas.

This book is amazing in its variety. From dangerous avalanches nearly destroying his family as a youth to survival and deaths of friends on other high mountains as an adult the experiences are vivid and riveting. From enduring the terror of the 9.2 1964 Good Friday Earthquake to climbing to the top of Denali and guiding numerous clients, the stories are compelling. Death is never far away. From being a member of the Alaska Rescue Group and rescuing numerous victims or recovering fatalities in the high mountains, whether from falls or avalanches, there is little time to take a breath.

And then to visualize and feel like you are there when hang gliders sail from the highest peaks including Denali and the inspiration that comes from contemplating those heights alone and mentally seeing them sailing free, thousands of feet in the air, is simply impossible to imagine. And to compliment this is a myriad of pictures that are extraordinary and incredible.

Jim educates us about each of these endeavors and shares stories from famous climbers, their feats and sadly, some of their deaths. He speaks of the draw, the adrenalin and the quirks. It is fascinating.

And there is the human element. Family, wife, and friends all bound together in their struggle to find themselves and build deeper bonds.

Jim Hale brings all of this and more together in a way that is touching, exciting, illuminating, and will bring to you a profound respect for Alaska and for all of those of us who challenge and treasure her.

"Tight Lines, Straight Shots!"
--Rocky McElveen, Owner of Alaskan
Adventures, big game guide, and author of *Wild Men, Wild Alaska* and *Wild Men, Wild Alaska II*

This book provides a glimpse of some of the exploits and adventures of Jim's life in Alaska. I came into his life during the phase of the high mountain season. His passion, zeal, and excitement for each adventure into the mountains were his *everything*. I preferred traipsing through

the valleys with the flowers and beaver dams to the cling-
ing to ropes on the high frozen heights.

Love, however, captured me, and I spent the early part
of our life following Jim up ice falls, sea-kayaking, and
climbing high altitude mountains. Climbing rock for an
afternoon jaunt was our idea of rest.

My best friend, husband, father of our three awesome children, and the love of my life, Jim and I have
been enjoying the challenge of *High Adventure* daily. The
title of this book, *An Alaskan Life of High Adventure,* has
two meanings. One is about the joy and terror of going
beyond the limits of one's self in exploration. The other is
the spiritual *High Adventure* that comes as revelation in
the midst of impossible situations by a living God who
watches over us.

Jim now leads the souls of men and women as a pastor
in surviving the extreme adventures of daily living.

It's been a wild ride being with him. The best choice I
ever made was to rope up with him for life!

--Roni Hale

INTRODUCTION

"Experience is what you get when you didn't get what you wanted."

-Anonymous

"We're going in," the pilot said in a calm, quiet voice.

"What do you mean we're going in?" I blurted out in response. He had forgotten the small plane didn't have the power at this altitude to climb out flying up-glacier. We searched the area, looking for a way out. There was none. The plane, with us in it, would crash on the crevasse-lined Ruth Glacier in a very long minute.

Adventure comes in many forms. There is the planned adventure; the moment of success that comes with good planning and good fortune where the calculated risk turns out as you had thought. And then, there is the unplanned adventure, the spontaneous, surprising type that can change a simple hike into a desperate and dangerous rescue at the turn of an ankle, the failure of your calculations or in overlooking some critical area.

This book is about successes and failures, the great experiences and the experiences that will teach you a lot if you live through them. There is an old saying in aviation that "any landing you walk away from is a good landing." These accounts are primarily about good landings, with the occasional fatal crash.

The book is about high adventure, which is, first of all, exciting and on the edge. Secondly, it's about the location, often in the air or on the great mountains, and then, it's about spirit and coming to know what lies beyond the natural.

Each of these experiences has left, in its wake, a new awareness of life and hopefully, practical wisdom. Much of my life and pastoral training came through these experiences; working with people in nervous situations, helping them attain their goals and sometimes, unfortunately, recovering the bodies.

There is a story told by marriage and family researcher Gary Smalley about a young bank president replacing an older, successful, and retiring president. The young man timidly came to the office of the retiring president and knocked on his door, saying "Sir, I know this is your last day on the job, and I hate to bother you, but I wonder if you might share with me what you feel to be the secret of your success." The old man growled, "Two words—good decisions."

"Thank you, sir," the young man replied. Walking down the hall, he said to himself, *Wait; that really didn't tell me anything.* He returned to the office of the old president and said "Sir, I'm sorry to trouble you again, but could you tell me how you made those good decisions?"

"One word," the old man said, "Experience."

"Oh, thank you sir, that will be very helpful," the new president said, but, as he walked away, he knew he still didn't have what he needed. He came back to the president's office once again. "Sir, I hate to bother you this third time, but how does a person get this experience?"

"Two words," the older man exclaimed, "bad decisions."

Wilderness adventure, like flying, is a matter of survival and learning from your occasional bad decisions. Experience that helps you make better decisions in the future is what you get, even if you don't reach your goal. This book is for all those willing to go beyond themselves to experience and to explore, who will try and try again until, hopefully they win through and reach their goals, whether they be spiritual, emotional, or physical.

BORN TO ADVENTURE

"Life is what happens to you while you're busy making other plans."

--John Lennon

Alaska is adventure. Everything about the place is extreme. Situated as it is, between Arctic and North Pacific weather influences, with its massive mountains, vast wilderness, and tremendous size, Alaska seems to never end in the opportunities it offers for exploration and new experiences.

My parents were adventurers. My father, Dr. George Hale, born and raised in Missouri, hunted, fished and floated rafts Tom Sawyer style through his boyhood years. Later, he was moved to a career as a medical missionary as he stood by the Egyptian grave of Bill Borden, a young man who gave up his privileged life for the sake of others and who died on the way to the mission field in China having not fulfilled his dream. He said to himself *I will complete your mission, Bill.*

He was an officer and ship's surgeon on the battle cruiser *U.S.S. Alaska* during the War of the Pacific in World War II. After the war ended, the *Alaska* was sent on patrol based out of Tsingtao, China. There, he came to the decision that he wanted to return to China as a medical missionary after his service to the navy ended.

My mother, Mary Hale, was an English and music teacher who also yearned to serve others with her life. Deeply spiritual, she had experienced as a young girl what she felt to be the hand of God come between her and a fast-moving train as it struck the car right at the passenger door where she was sitting, saving her life.

Their desire to serve had to wait however, as shortly after the end of the war the communist Chinese began overrunning China and killing missionaries. The mission board said they should reconsider unless they felt "called to be martyrs." They were left without a plan. They didn't know what to do.

Alaska was the solution to the problem. While working in Washington, D.C., at Washington General Hospital, Dad performed a surgery for an official of the Alaska Railroad, who talked him into coming to the Territory of Alaska as Assistant Chief Surgeon for the Alaska Railroad.

After World War II, the Territory of Alaska was primarily made up of military bases and a few small cities and villages, with little other development. The people of the Territory were looking for professionals of all kinds to come and help build the newly emerging state. The wilderness and adventure of Alaska were enticing. Dad and Mom arrived at Elmendorf Air Base, the only large air field in the area, in 1949. Anchorage had approximately thirty-five thousand residents; the whole state's population was a bit over one hundred thousand persons. Dad would be the first civilian surgeon in Anchorage. In those days, he did everything from delivering babies to brain surgery, as it was so difficult and time consuming to send the patient outside to Seattle or somewhere else in the lower forty-eight states.

Dad loved to hunt and fish, became a private pilot and flew what he called "The World's Greatest Super Cub." He lived through plane crashes, avalanches, and earthquakes. He was a great story teller and jokester his whole life. For example, in chemistry class at William Jewell College he discovered a compound that would stay clear until it became wet. After becoming wet it would turn bright red. He thought to himself, *There must be an interesting use for this somewhere.* He decided the best use of his compound would be to paint every toilet seat in his dormitory but one with the chemical. Later, in the showers, bright red rear-ends abounded with no one the wiser as to how or who.

In his retirement years, he would set dozens of Alaska state age group records in swimming and track and field events and keep written notes on the lives of the friends with whom he trained, in order to remember them well. His adventurous spirit was demonstrated when, in his later years, he and Mom visited my brother John and his wife, Deena, at a wilderness lodge they were developing at Warm Springs Bay on Baranof Island. Dad surprised them all by jumping into a canoe to try and chase down a pod of killer whales that had come into the bay. He didn't lack for adventurous spirit!

Story after story comes to mind that demonstrates the adventurous spirit of my family and of Alaskans in general. Here are a few from the territorial and early statehood years.

Alyeska Ski Resort, a world-class ski area south of Anchorage, had not yet been developed in 1958. A small Bell helicopter, piloted by Link Luckett, shuttled skiers for five dollars a trip to the ridge beside the glacier on the upper mountain. Dad, my brother John, and I took part in those early trips. I was five years old. Alyeska is a steep and challenging mountain with 2,500 vertical feet of terrain. My skiing skills were limited to what is now called the Wedge. In my red jacket and Davy Crockett coonskin cap, I did my best to keep up with the group. No one could express the pain involved with holding a wedge-shaped snow plow down the entire mountain on an icy spring crust!

Dad had his own ideas about what we could handle. He would often set me and my Winchester .243 alone, at six years of age, on the bank of a pond in bear territory to wait for a caribou to come for a drink. I still marvel at his expectations and that we managed to survive them. Looking back, I see that I passed many of these too-high expectations on to my own children, though I hoped not to.

In 1963, bush pilot and guide Dennis Branham flew Dad and my older brother, John, then twelve years old, to the top of a mountain in the Alaska Range for a ski descent that nearly destroyed our family. I'll let Dad tell the story in his own words:

"In March of 1963, Dennis and Millie Branham invited the Hale family to spend a weekend at Rainy Pass Lodge on Puntilla Lake in the Alaska Range to explore the possibilities of skiing in the area.

"Mary and I, with Johnny, age twelve, Jimmy, age nine and Nancy, age six, were flown the 140 miles to the lodge in two ski-equipped airplanes, a four place Helio-Courier, which is a very high-performance, short take off and landing plane, and a four place Taylor Craft piloted respectively by Dennis and Jimmy Branham. On a beautiful Friday afternoon, we flew by 20,320-foot Mt. McKinley and many other beautiful, rugged, snow covered mountains on the way. The next day, while John and Jim hunted ptarmigan and snowshoe rabbits on snowshoes with the son of the lodge caretaker, Dennis and I loaded my skis and poles into the Helio-Courier, took off, and flew around the mountains bordering the broad Ptarmigan Valley. Above the lodge, we searched for mountains with tops flat enough to land on, good slopes for skiing and a landable area near the bottom. "Eventually, we landed in a small bowl at 5,400 feet near the top of a mountain about three miles from the lodge.

"We unloaded my skis and poles, and after Dennis had taken off and climbed away without mishap, I climbed, on skis, to the 6,000-foot peak of the mountain above the bowl where we had landed.

"Then, I skied down into a bowl and down a ravine running toward the lodge.

"Soon, I came to a steep, narrow, seven-foot-wide chute bounded by vertical walls of sharp rocks. As I skied down the chute, the tips and tails of my skis scraped the rocks at the sides.

"Emerging below, I stopped and looked back, hoping to find a better course around the chute.

"As I looked up the mountain, I saw a long, steep ravine to my left filled with smooth powder snow, which led into the ravine where I was standing. There was a large cliff between the two ski runs but, above the cliff, was a large shoulder, like a ridge covered by smooth, wind-packed snow. I decided that if I should ski this slope again, I would traverse across that shoulder and ski down that beautiful powder snow instead of scraping my skis on the rocks forming the sides of the chute.

"I then skied on down the ravine, worked my way through a tangle of alders in the creek bottom, skirted the ridge to the left, and skied down a hill and across the frozen snow of Puntilla Lake, arriving at the lodge.

"Sunday morning the temperature was seventeen degrees, partly sunny and slightly overcast. Johnny and I planned to ski together. Jimmy preferred to hunt with the caretaker's son again, and I encouraged this, not wanting all of the male members of the family to be landing on the mountains in the same airplane. This was fortunate since I don't believe a boy of nine could have survived the day.

"After breakfast, Johnny and I loaded the plane, took off and flew around the mountains again, looking for another good run to ski. The air was rough and we could find no other satisfactory mountain, so, eventually, we decided to try the same run I had skied the day before.

"Since there were some rocks protruding through the snow where we had landed, Dennis wanted to land in his tracks of the previous day, not having scraped the plane's skis on rocks in that landing or take off.

"We approached the mountain bowl about a dozen times but kept going around due to the turbulence and the overcast, resulting in a whiteout condition that caused the previous day's tracks not to be seen in time to land smoothly. Johnny was almost airsick when we finally saw the tracks early enough in the approach for Dennis to land.

"We unloaded our equipment, put on our skis, took some movies and began our descent down the gorge. Dennis had a movie camera and took pictures of us from above as we began our descent. We had another camera with us and took pictures of each other occasionally.

"We began to warm up, seemingly from the exertion, and took off our outer jackets and put them into a rucksack, to keep from getting wet with sweat.

"When we were just above the narrow chute I had skied the previous day we turned right and began our traverse across the steep shoulder as I had planned.

"Suddenly we heard a loud crack above us. We looked up to see what was coming down on us and immediately discovered that the entire area of the shoulder, which had been covered by an 18" slab of wind-packed snow, was sliding fast, with us in the middle of it!

"We plummeted down a very steep, sixty-degree slope, sliding with the snow between sharp rocks projecting three to four feet above the snow, and dropped over the brink of a cliff. We fell through space, completely blinded by the falling snow, for what seemed like an hour. Finally, I wondered, *Aren't we ever going to hit?*

"Just then we crashed, glancing off a steep bank of crusted snow. I hit on my right back side, feeling a sharp snap in the ribs of the left front part of my chest and in my lower back. My ribcage had broken open like a dropped watermelon and my lower back was injured.

"We floated down a steep gorge in the flowing avalanche. I found myself swimming instinctively and desperately with long powerful surfacing strokes, trying to keep my head above the snow. Large blocks of snow were sailing swiftly and silently down the gorge and over my head, confirming other observations that avalanches travel like a caterpillar tread, faster on top and slower on the bottom, due to surface friction."Suddenly, and mercifully, the avalanche stopped. My head was above the surface of the snow. The rest of me was buried in a shallow grave of snow. I stood up, throwing off the snow, and was free.

"My ski bindings had released and the safety straps attaching them to my boots had broken in the fall. My ski pole straps were still around my wrists. I looked around and down the gorge to find I was entirely alone. There was no sign or sound of Johnny. He had vanished entirely!

"I said, 'Oh, God, No!' I was badly injured and the gorge was filled with many tons of snow. Statistically, fifty percent of people buried in avalanches are dead in thirty minutes, ninety seven percent in an hour—so there was no time to spare!

"I had no idea where he was, but he had been below me when the avalanche started, so I walked down the surface of the avalanche debris, feeling and hearing my broken bones grinding with a very unpleasant sound like a loud 'whump' with every step. Mercifully, I was still too numb to feel the pain. Suddenly, I heard a voice from under the snow say, 'Get me out of here fast!' I picked up a large block of crusted snow, (about 2 feet square by 18 inches thick) at random, and found it was right on top of his racing-helmeted head.

"He was completely locked in the snow and was able to breathe and talk but, otherwise, was not able to move even a finger. While carefully excavating him, I discovered he had apparently been knocked out by the fall, remembering only the loud crack at the start of the avalanche. Waking up being unable to see or move, he heard the crunch of my steps down the avalanche surface above him and yelled. Probably his racing helmet saved his life.

"After Johnny was released, he seemed to have only a sprained ankle and thumb.

"He commented, 'This makes you feel a little shaky, doesn't it?'

"I heartily agreed.

"I started to walk after getting him out, but with each step my foot would break through and sink about nine inches into the snow and my ribs would dislocate with a loud grinding. I found myself completely immobilized where I stood.

"Johnny took the movie camera and made a quick sequence of the fault remaining above the shoulder over the cliff where the slab of snow had separated, the approximately 50-foot cliff, the surface of the avalanche with the shallow grave where I had stopped with my skis being visible beneath the snow, the place where he had been entombed, and me, standing and dumping the snow out of my pockets. Needless to say, it had penetrated everywhere, filling my clothes and my pockets, too.

"Then Johnny climbed up the avalanche, retrieved my skis and put them on for me since I was almost helpless. I noticed that my sunglasses were missing, and Johnny made another trip up the avalanche and found them buried in the snow where I had surfaced.

"With skis in place, I stayed on top of the snow instead of sinking into it. My broken bones displaced much less frequently, and I was able to travel again.

"We left as soon as possible since I wanted to get as far as I could before the blessed numbness wore off of my fractures.

"We skied on down the ravine, and I stemmed as much as possible and lightened my skis as little as possible in trying to avoid displacing my fractured bones. By the time we were down to the tangle of alders in the creek bottom, I began to feel pain and couldn't suppress a few loud groans when my skis became entangled, resulting in falls.

"Since skiing is a one-person operation, Johnny could not help at all, so I told him to ski back to the lodge and tell the others what had happened and that I was hurt but could make it back in.

"By the time Johnny arrived at the lodge, Dennis had become concerned over the delay and was about to fly out and look for us.

"I eventually worked my way through the tangle of alders, climbed over the ridge, skied down the hill, crossed the frozen lake, and arrived at the lodge. There, we learned that the temperatures had risen dramatically during the day, from seventeen degrees to forty-eight degrees in a short time, probably contributing to the weakening of the snow and the resulting avalanche triggered by our weight.

"I had my friends tie a triangular bandage tightly around my chest and took a couple of stiff, pain-killing drinks of whiskey. Dennis and Jim then flew us back to Anchorage.

"After X-rays at Providence Hospital indicated I had at least three fractured ribs and a fracture of the cartilage holding the rib cage together, we went home. Later X-rays would show a fracture dislocation of my lower back. The rib injuries had been so painful that the lower back injury had been over-looked for a while.

"The next morning, in spite of medicines, I woke up in pain at 3:30 a.m. Thinking that I might be in less pain sitting up, I tried for half an hour to get up by myself in order not to interrupt my wife's peaceful sleep.

"No matter how I turned or maneuvered, I couldn't get up. Finally, I had to awaken my wife, Mary Helen, and ask for a lift. For the rest of that week I couldn't get out of bed without help but, once up, was able to get around.

"That Monday I had an eight o'clock operation scheduled, which I performed. But, for obvious reasons, I didn't help lift the patient from the operating table to the stretcher after the procedure was completed.

"At noon, as was my custom, I walked the seven blocks downtown to the Chamber of Commerce luncheon.

"After working that week, seeing only very essential problems, I had the next week at Mt. Alyeska as Chief Surgeon for the National Alpine Skiing Championship. Skiing cautiously a little was tempting but, after taking one ride up the main chairlift and finding that it hurt when the chair bumped over the tower rollers, I decided against it.

"The next week I resumed full practice. One month after the avalanche I was in a sporting goods store, by which I had passed while walking downtown the day after the accident. One of the employees, who had worked there four years previously and had just returned, said, 'I heard you broke some bones a month ago.' I verified that such was the case.

He explained, 'I had just returned to Anchorage after an absence of four years, saw you walk by slowly, all hunched over, looking pale and grim, and said, 'Dr. Hale sure has aged in the last four years!'

"Four weeks after the avalanche I was again skiing, cautiously but enthusiastically, at Mt. Alyeska.

"Johnny was not only skiing wildly, as usual, but spent a beautiful, warm, sunny afternoon skiing down the south face of the mountain, across a valley floor, up a sharp ridge, sailing 180 feet through the air. He would pass by fifteen to twenty feet over my head with his clothes snapping in the wind. What a recovery!"

Our family was always on the move. If it wasn't skiing or running rivers in Dad's specially modified canoe we would be playing baseball, bowling, dip-netting for Ueligan (Smelt) fish, or a myriad of other activities. Dad loved to float the icy glacial rivers of south central Alaska. He took a long, flat-bottomed lake canoe and built a special spray skirt for it. The canoe looked more like an aircraft carrier than a kayak. Together with a battery powered bilge pump and our wet-suit jackets we floated and explored many a whitewater river. Our lives were full and happy.

Mom was not quite so adventurous, but tagged along on the hunting and ski trips. After injuring her shoulder skiing at Arctic Valley, she began to focus more on home life, gardening and the community of Anchorage, Alaska's largest city. She was founder of the Alaska

Festival of Music and one of the early conductors of the Anchorage Community Chorus. She was so influential in her passion for the arts and fundraising skills that she received a plaque from friends that said, *Before you call Washington, call Mary Hale!* Nominated many times for the *Alaskan of the Year* award she was a "never give up, never say die" type of person, as well as a brave and deeply inspirational woman.

On a hunting trip in the late '50s she had an opportunity to shoot a caribou. Dad had shot another caribou over a mile from the camp and was packing the meat to the lake, to be flown out later. While he was away, a caribou walked right into camp and Mom shot and killed it with her Winchester .30-.30. Being very tender-hearted, Mom cried and swore she would never shoot another animal again. Dad was upset when he found, after a very long and exhausting day, that he had another animal to dress out and pack to the lake.

Life in Territorial Alaska was a dream for those who enjoyed the wilderness. There were so few people. Hunting and fishing rules were very relaxed in comparison to what they are today. For example, you could fly and hunt on the same day. Dad worked as a hunting guide for Bud and Dennis Branham of Rainy Pass Lodge on his vacations and had a great reputation for strength and skill in the bush.

Our young family spent much time in small planes on floats, wheels, or skis, spending Christmas at Rainy Pass Lodge or fishing with the Branhams in the summers. Unfortunately, children are often susceptible to motion-sickness. Flying in small bush planes can be a wild ride at times, and we threw up in Dennis's hat more than once.

I grew up in hunting camps, plinking away at cans and caribou guts with my modified BB rifle and shot my first caribou with a Winchester .243 at seven years of age. I so enjoyed being out that I can remember to this day, many years later, the sights and smells of the fall tundra turning orange-brown after the first frosts, the sweetness of frost-kissed blueberries, and the taste of a lunch that we shared as we glassed the rolling tundra around us with binoculars, looking for animals.

The Territory of Alaska became the State of Alaska in 1959; I was six years old at the time. The population was growing and the state was coming into its own. The year of 1964 would test whether we had it in us to survive, as individuals or as a state.

The Alaskan Good Friday earthquake was one of the largest ever recorded at 9.2 on the 10 point Richter scale. Entire villages were left high and dry as large areas experienced massive uplift when the tectonic plates of earth collided. Many were wiped from the map by powerful tsunami waves. Entire towns, such as Valdez, had to be relocated as they dropped into the reach of the ocean tides.

My mother and I were in the kitchen of our home overlooking Cook Inlet when it began, and my sister Nancy was in the basement on roller skates. Earthquakes were common and I remember laughing and saying, "Ha, ha, earthquake." Suddenly the quake struck in earnest with a paralyzing and ferocious shock, and the laughing stopped.

Not knowing where she was, we attempted to find Nancy, calling out her name, stumbling through the house, being thrown violently from wall to wall like a pinball in a pinball machine. The earth quaked for over three minutes. The roaring noise of the quake and the house being torn apart was indescribable. Our house was broken into pieces as the primary fault line that went through downtown Anchorage destroyed it on the way to taking out Turnagain Heights, where seventy-five homes were destroyed as they slipped down the bluff into Cook Inlet. Finally, after what seemed like an eternity, the earth stopped its rolling and shaking. Nancy and her friend, on roller skates, had taken refuge under the bar just as a large mirror shattered and showered the room with broken glass. They were unharmed.

Dad tried to reach us to see if we were all right. He was unable to drive through the shattered streets and tried to run home from his office, only to be stopped by a large crevice just short of the house. Talking from opposite sides of the crevice, we assured him we were okay, and he headed off to the hospital to spend many difficult days taking care of the injured.

What was left of our house was unfit to live in. It had been broken into three major pieces. Gas lines were ruptured, and there was a real fear of fire. Our living room floor tilted at a strange angle. Police and state trooper units traveled the streets with loudspeakers, warning people of a potential tsunami and telling them to head for high ground. We spent the night huddled in our small Corvair automobile waiting out the anticipated wave. Many lives were lost in Seward, Valdez, and other small villages as the tsunami hit. Thankfully, in Anchorage, there was no significant wave.

It's hard to comprehend as a ten-year-old. I lost a friend, Perry Mead, Jr., in Turnagain Heights, as he went back into his home for his baby brother during the quake and was never seen again as the house collapsed and slid down the bluff. Many of the homes in Turnagain were destroyed as the frozen ground they were built on slipped on a layer of glacier-deposited blue clay toward Cook Inlet. The silty clay turned to jelly as it was shaken and put under pressure by the quake. At the same time, I remember playing with boats in a puddle with a friend and remarking to each other how cool it was to live in a disaster area since school was closed for two whole weeks.

We tried to move back into our home, but a contractor friend of my father's talked some sense into him after a large aftershock earthquake one week later. Frightened by the powerful aftershock and the memory of the Friday before, we had run out of the house and jumped over a large crack just below the front steps. Our friend asked Dad, "Dr. Hale, come with me if you would. There's something I need to show you." He made his point by extending his thirty-foot steel tape measure down into the crack in the earth and swinging it back and forth without touching anything.

We left. The house was later sold for fifty dollars, chain sawed into pieces and moved.

Our lives were changed forever. Very few people in Alaska had earthquake insurance. Our home was destroyed, and even the land itself was unstable. We had to start over. Both Dad and Mom went into emergency mode and stayed there so long they forgot how to rest. It took my overwhelmed parents many years to finally come to a place where they felt at home and at peace again.

We looked for a new home somewhere far from the unstable ground near the water. A new stage of life began in a new neighborhood in South Anchorage, in the foothills of the Chugach Mountains. We had new schools, new friends, and a new life in an area with many moose and the occasional grizzly bear. I would spend many wonderful hours exploring, making trails on snowshoes and skis. I just loved seeing around the corner or finding new vistas at the top of a ridge or mountain.

Our family had a small Cub Cadet tractor lawn-mower. It seemed like a good idea, in my ten-year-old mind, to drive it up into the mountains, so the Muckey brothers, Greg, Steve, and Jeff, hopped in the trailer and we took off for a high mountain lake in the DeArmoun Road Valley, many rugged miles away. We had found a gas can and brought extra gas, lunch, and a rope as safety gear. Assuming the liquid inside the can was gasoline, I didn't bother to check when I filled it up.

After lunch well up the valley and high in the mountains, we filled the tractor's tank from the gas can. Not long after, the engine stopped—dead—and we were forced to haul the tractor many grueling miles with our safety rope. We must have looked like a very strange sled-dog team. Finally, after miles of hauling the tractor uphill and coasting downhill a man stopped his pick-up truck alongside and asked, "Need any help?" Did we, ever! Checking the basic functions of the engine, he finally came to the carburetor, emptied out the sediment bowl and told us, "This is full of water." Draining the water from the fuel tank, we started the engine and cruised into the yard as if nothing had happened. I still don't remember ever telling my parents about our exploratory drive into the Chugach Mountains.

My explorations took on new depths as I became a certified scuba diver at thirteen. Seeing the new sights of the undersea world, the colors and the strange creatures increased my appetite for new experiences. Alaska had so much to offer as an experiential playground.

I was being introduced to the mountains, which would later consume my young adult life. Good high school friends, Robbie and Jimmie Johnson, had begun rappelling the rock cliffs along Turnagain Arm south of Anchorage and invited me along. It was radically exciting as we slid down the rope in long, sweeping jumps over one-hundred-foot cliffs. It seemed natural to want to find new ways back up the cliffs, and so we began rock climbing. The challenge and excitement caught me. I was hooked.

Each year on the Fourth of July, hundreds of assorted "crazy people" gather in Seward, Alaska, to run the famous and popular Mt. Marathon Race. Thousands come to watch. Begun as a bet in 1908, the race was officially established in 1915 and is considered one of the oldest annual footraces in American history. The original bet was that a runner could go from downtown Seward to the top of 3,022-foot Mt. Marathon and back within an hour. Drinks for the crowd were the stakes and it was decided to run the race on Independence Day to add to the festivities.

Today, the record time is owned by Alaskan Bill Spencer, who completed the round trip in forty-three minutes and twenty-three seconds. The race attracts adventurous runners from all over the world and is quite a spectacle. Though the ascending speed seems turtle slow at two miles per hour, the average speed on descent is a breakneck twelve miles per hour as the runners cover over twenty feet per stride down the soft scree-filled gullies.

In 1969, while training for the equally popular junior version of the race up Mt. Marathon, I met God. A friend and great high school runner, Dan Mardock, and I were rooming together in the old Alaska Hotel, owned by a Mr. Hit. Mr. Hit's son, Dennis, had died in a fall while training for the race up the mountain a few years before and Mr. Hit, as a memorial, provided room and board for any runner training for the race.

Dan had a small, hand-inscribed pendant he wore on a chain around his neck that said, "Trust God." In typical adolescent fashion, I was ridiculing his trust in a God I didn't believe in. The turbulent '60s with all the cynicism and hate had affected me deeply. I considered myself an atheist, but now recognize I was angry at the God I thought should be there and seemed not to be. In angry mockery of his simple faith, I tried to start an argument, spitefully saying, "Mardock, if your God is so real, why doesn't he just kill me right now and prove it to you?"

Dan replied, "Hale, I feel sorry for you. You have no idea how much God loves you."

Right at that moment, I felt God's presence in front of me, as real as my heartbeat. I have never lost the feeling of that moment. Suddenly, I knew there was more to life than I had understood. God's presence as a companion and friend would literally save my life in coming years.

After graduation from A.J. Dimond High School in 1971, I attended two years of college in Gresham, Oregon. As a competitive runner and cross-country skier, my life had been focused on school and athletic competition. Having reasonable success at the junior college level, it seemed logical to continue the athletic pursuits, but something was changing within me. Dad had bought a book that intrigued me, *On Snow and Rock*, by Gaston Rebuffat, a mountain guide from Chamonix, France. My dreams began to change as I read the stories and saw the photos of the high Alps. I was questioning my future, and the mountains, with all their rugged beauty and promise of adventure, were filling my heart.

During December of 1973 it became necessary to make a choice; the proverbial fork in the road lay before me. Should I continue in school and athletics, hoping to fulfill childhood Olympic dreams, or should I pursue a career as a professional mountain guide? Leaving the Glen Alps parking lot below Flattop Mountain above Anchorage in a windstorm, with a question, I walked in the wind-driven snow, looking for an answer. Somehow, I returned to that parking lot hours later knowing what I would do. Though the odds of it happening were not very good in the United States of 1973, I would focus all my efforts and energy on the goal of becoming a professional mountain guide.

THE CALL OF
THE MOUNTAINS

"I've stood in some mighty-mouthed hollow
That's plumb full of hush to the brim."
--*The Spell of the Yukon* by Robert Service

My life as a young man was noisy. In my family, a TV was always going somewhere, radio alarm clocks would wake me with music, which I played constantly whether at home or in the car. My life was active with school, athletics, and friends. There was little time for quiet. Occasionally, I would get out by myself and snowshoe or hike and feel the quiet, but mostly my mind was occupied and filled with constant distraction.

As I explored the Alaskan mountains and wilderness, quiet began to penetrate my soul. Both exhilaration and peace became tangible. I felt more alive than I had ever been. This sense of meaning became purpose. I couldn't get enough of the mountain environment.

Mt. Spurr is a large volcano visible from Anchorage to the west on a clear day. The southwest ridge of that mountain was beautiful from a distance, sweeping up over ten thousand feet and looking over the crater of the active volcano. Every two decades or so, the mountain would awaken and spew volcanic ash over the area, closing airports and being a general nuisance. My friend, Jim Johnson, and I decided we would climb it during the summer of 1973.

Trying to find a way to the ridge was almost as far as we got. I did a solo reconnaissance a few weeks before the date of our climb. Flying into Chakachamna Lake with my father in his float equipped Super Cub, I thought I could see a creek drainage that would lead us onto the ridge.

Brush, specifically alder brush, can present a terrible difficulty in backcountry traveling throughout Alaska. There are many stories of climbers and hunters alder bashing to complete frustration in an attempt to get somewhere. On top of that is the ever present Devil's Club, a thorny and painful nuisance often found in the midst of alder patches.

My exploratory hike on the glacier toward the creek bed was simple and easy. It was encouraging to travel so many easy miles. But then came the alders. Coming down off a lateral moraine, the gravelly rubble piled up alongside a moving glacier, I entered a large and unavoidable section of alder. Only two hundred yards away, the creek bed looked easy to reach. An hour later, and in exhausted frustration, I had given up and retraced my steps to the glacier's edge. The trees were so close together and leaning downhill in a sweeping arc that it was impossible to travel without a chainsaw. Sweaty, tired, and discouraged, I happened to notice an area where a large rock fall from the glacial moraine had broken through the alders and cleared a path. *I wonder*, I thought to myself. Twenty minutes later, I was at the creek, happy and sure of our approach path.

Knowing now that Jim and I could actually get to the mountain, I cached some fuel for our mountain stove and some bamboo wands that we would use as trail markers on the snowy, upper mountain under an overhanging rock. I waited a short time at a tiny, old cabin on the shore of Chakachamna Lake and returned to Anchorage with Dad in the Super Cub.

Three weeks later, we arrived with our full and very heavy expedition packs to begin the climb. We moved through the alders, up the creek bed, and onto the ridge in one long day, camping at nearly 4,000 feet on the last of the soft tundra vegetation before the more difficult climbing began.

It's typical in Alaska to be surprised by the conditions on the ground once you get there. We started off the next morning with great hopes of camping well up the Southwest Ridge. We soon ran into deep volcanic ash that sapped our strength and shattered basalt rock ridges that made it extremely difficult and dangerous to climb with our heavy gear. Struggling through the deep ash, wearing our heavy packs, we said to each another, "This is ridiculous." Trying to find our way over the steep and broken basalt ridges we finally said, "Enough. This is too dangerous with these packs." With real disappointment, we backtracked to our camp of the night before.

After a night's sleep, we decided to try a "light and fast" climb to the 10,000-foot first peak of the Southwest Ridge. It was much easier moving through the ribs of broken basalt with light day packs, and we were soon on the upper mountain. My father flew by in the Super Cub at 9,000 feet to check on us. Having no radios, we attempted to communicate with him that we were stalled and would be back at the lake the next evening and to pick us up. Messages stamped out in the snow can be difficult to read. He didn't understand we would be descending.

Finishing the climb on a brilliant, blue sky afternoon, we were treated to expansive views that covered hundreds of miles. A cloud layer lay in the valley many thousands of feet below. It was entrancing and unearthly. We were in another world.

We came off the mountain very quickly, laughing and shouting with glee as we glissaded down thousands of feet of snow slopes in the high mountain beauty. Packing up our camp, we continued to carefully descend, with our heavy packs, into the valley below. Coming down into the creek bed, Jim stepped on a large, sloping wet rock. I heard a quick gasp as he slipped and began a headfirst, sliding fall into the creek bed, thirty feet below. My heart was in my throat as I watched him pile into the boulders below. "Jim, are you okay?" I yelled as I quickly but carefully descended to his location, helping him out from under his pack.

Holding his wrist, sprained in the impact, he replied, "Yeah, I think so, except for spraining my wrist." Taping the damaged wrist, we took a deep breath, thankful the incident didn't turn out much worse. Happy, chastened, and tired, we moved into the small cabin near Chakachamna Lake late in the evening, expecting to be picked up in the morning.

We waited four days. Much of the time was spent laying around on small, rope beds or walking on the beach. We were awakened one morning by the sounds of scratching on the cabin wall. As we looked out the one window, something black appeared in the brush that looked like a porcupine. *No big deal*, we thought.

A few minutes later, as Jim sat on the bunk in his underwear, with his back to the window, I barely had time to react as a large, black form flew toward him. The black bear hit the window sill with its paws, making a loud bang. I thought Jim would hit the roof as he came off the bunk. "Get out of here!" we screamed loudly as we banged cooking pots together in an attempt to frighten off the bear. With only our ice axes as weapons we hoped the bear wouldn't try to come through the window.

Thankfully the bear went on its way. The cabin door had been fitted with "bear boards," which are boards with nails driven through them from the other side and then attached to the door, making a sharp and uninviting surface. The window was off the ground about five feet, making it a difficult crawl for a bear attempting to climb in. Not that they didn't think about it. Shortly after this intrusion, a large, male black bear with a broad head much like a grizzly bear's stopped by to inspect the cabin.

Standing on his hind feet, he pawed at the glass with great delicacy, seemingly amazed at the glass itself.

We were scared. The cabin was in a patch of alder brush fifty feet or so from the open beach, where we could see whatever was coming. Curious bears kept moving by the cabin—our socks kept disappearing from the drying line. Due to the brush, we had no way of knowing if a bear was only a few feet away as we opened the door to leave the cabin. We began a ritual of opening the door and quickly pulling ourselves onto the roof to search the brush for anything of concern.

Finally, the plane arrived, but the waves on the lake were dangerously high and, after loading Jim and his gear, my father told me, "Jimmy, I can't come back to this beach in these waves. I need you to hike over to the mouth of the Chackachamna River where it's out of the wind." Being alone and without a weapon, it felt like a longer walk than you might measure on a map. I was really concerned about bears and constantly looked around as I hiked the glacier rubble to the lake's outlet. Luckily, I had no bear encounters.

It was great to be home in Anchorage. It felt safe, peaceful, and familiar. But something inside me had changed. I had experienced an awesome simplicity. I considered all this as I sat in Dad's car, watching the telephone poles and houses go by, amazed that I was moving without making any effort.

SOARING WITH THE RAVENS

I enjoyed being alone in the wilderness. It was beautiful, exhilarating, and quiet. Within myself, I wondered at the power of its effect on me. Some of its allure was the wonderful feeling of peace and freedom found in each experience, some because of an insecure desire to be away from others' scrutiny, correction, and control. The magnificent beauty of the mountains and the ocean were worth the effort. I loved exploring, hiking, climbing, and diving alone, though it was always a risk. Should something go wrong in the days before cell phones, GPS, and personal locator beacons, you were truly alone and had to take care of yourself in an emergency. I developed an attitude of being an "adventurous chicken." I loved to explore, but also had experienced that injury and the resulting pain hurts and I would do all I could to avoid it by being extra careful and attentive. One of the great moments in my life came as I soared with the ravens in my hang glider, alone in the Chugach Mountains.

I was fairly new to hang gliding in 1975 and had just purchased my own kite, a blue, white, and orange Seagull I. Carrying it up Blueberry Hill below Flattop Mountain, I was planning a short, low to the ground practice flight. As I set up the kite, the winter wind began to blow more strongly from the north, making the small hill a better flight opportunity.

Ravens began to soar and play along the ridge. No other bird plays like a raven. They will soar for hours, in perfect control, turning upside down, doing half-rolls and full barrel rolls; they're incredibly fun to watch. Hang gliders dream of soaring and playing with them in the uplifting winds.

As I launched, the winds were steady and perfect. I climbed a short distance above the ridge and, for a few minutes, was an accepted member of the group as the ravens soared together with me. I must have looked to them like one big, strange, multi-colored bird. They weren't bothered at all with my presence. It was magical as we flew together in the winter light of the late after-noon, looking out over the Anchorage Bowl below.

As I ground-skimmed through the low scrub pines into the Glen Alps parking lot, I did a quick, ground-level turn and flared the kite, dropping to a stop in the deep snow next to the road. I felt just like a bird. Though I would excitedly try to explain to my friends, I would never be able to communicate with words the joy of being like one of the ravens for that short but special time.

TRAPPED ON MT. HOOD

My love for the mountains was constantly increasing. While attending college in Oregon, I took every oppor-tunity to be out rock climbing or in the mountains. Even in the rain, alone, I found great joy and significance in just being there.

One winter weekend my roommate, John Adamson, a friend, Woody Krugel, and I decided to climb Mt. Hood by the normal Hogback route above Timberline Lodge. The common Northwest cloud cover was blowing up the mountain, but we were able to do the first mile easily enough by following the Miracle Mile ski lift to its end, well above tree line.

From the top of the lift, it was a pure whiteout, but occasional glimpses of the sun made us think the cloud cover was thin and we might break through it. So, off we went.

After a couple of hours, we found ourselves climbing onto progressively steeper slopes, with no certainty of where we were. It was too dangerous to continue. Using our compass, we began the descent. With the wind now in our faces, the impenetrable cloud coupled with sloping terrain caused us to wander off-track in spite of our compass. After an hour or so, we stopped. I asked John and Woody, "Do you have any sense of where we are?" We should have been nearing Timberline Resort but saw no trace of it. We lost faith in my compass work and became entirely disoriented. "Let's just keep moving down," I said as we continued the blind descent.

Suddenly, we came to an unexpected series of high cliffs. "Do you remember seeing anything like this above the lodge?" Woody asked. We were significantly lost. Climbing and descending in the thick cloud had us so turned around we were completely unsure of our location. The decision was made. "We need to dig in and wait until we have enough of a view to see where we are."

We began to dig an emergency snow cave using a small shovel and a cooking pot to hollow out our temporary home. In a small cave, with skis, ropes, and packs to insulate beneath us and a stove to melt snow for water we would be fine for a while.

Just as the cave became large enough to climb into, I felt I was nearly torn in two by a severe hamstring cramp. Screaming in pain is problematical when it hurts too much to breathe. "Aahh! Aahh!" I blurted out as I crammed my head into one end of the too-short cave and my feet into the other wall in an attempt to straighten the leg.

"What's wrong?" John and Woody yelled through the small cave entrance.

Breathlessly, I gasped, "I've got a cramp in my leg, I'm stuck."

John crawled into the cave entrance, reaching in to massage my leg saying, "Take it easy, you're going to be okay, relax." It seemed an eternity but after a few minutes the knotted muscle began to relax and I could take the risk of carefully trying to bend it enough to slide out of the cave.

Finally, as full darkness came upon us, we crawled through the entrance into a snow cave that was just large enough for the three of us to lie flat, shoulder to shoulder, like so many sardines. Lying on our packs, rope and other gear, we were well insulated from the snow and reasonably warm. We wondered how long we might be in this situation. We waited, talked, got tired of talking, and waited some more. Sleep was difficult and, worse, I had gone to see a new horror movie with my roommates the night before. Whenever I closed my eyes, I would see a portion of the movie in my mind's eye where a demon-possessed little girl's head would spin around. It certainly wasn't peaceful where I was lying.

Around 3:00 a.m., John went out of the cave to relieve himself and yelled back to us, "Hey guys, they've got the lights on!" Scrambling out of the cave to the beautiful sight of Timberline Lodge and the lights of the Miracle Mile chairlift glowing in the blackness below Woody shouted, "Man, that looks so good!" Scoping out our return route, I said, "Okay, we need to climb up and right before we angle down to hit the chairlift." We were silenced again as the clouds drifted in but happily returned to the cave knowing where we were. The next morning, still in the clouds, we climbed up away from the cliffs of Mississippi Head and descended back to the top of the ski lift, from where it was an easy ski back to Timberline Lodge.

We were a happy, humbled, and relieved crew as we debriefed with the Forest Service Ranger who was waiting at the lodge to begin mounting a rescue if we had not made it back on our own. That's when I decided I'd stick to comedies and dramas—the horror films were out.

ALASKAN MOUNTAIN GUIDE

"Twenty years from now you will be more disappointed by the things you didn't do than by the things you did. So, throw off the bowlines, sail away from the safe harbor, catch the trade winds in your sails. Explore, dream, discover."

--Mark Twain

I did all I could to gain the experience necessary to become a professional mountain guide. The mountains were everything to me, my source of meaning and purpose and the place where I found the greatest peace I had ever known. My greatest desire was to find a way to live and work in the midst of the snow and ice giants of the world. While still in college, I began attending Outward Bound and expedition schools from the Pacific Northwest guide services such as Rainier Mountaineering on Mt. Rainier, Washington.

Winter experiences in the Cascade Mountains taught me much about hypothermia, frostbite, team dynamics, and most importantly, about myself. I began to see with new eyes as Outward Bound School teammates would fearlessly and honestly debrief the trips and evaluate the personal dynamics. I had much to learn about working with people.

Much of my concept of leadership had come from peer and coaching experiences. It was more of a "Do what I tell you," and less of a learning experience. As I saw myself through the eyes of the other students in the Outward Bound Winter School, I had an epiphany. Leadership was less about directing and more about helping others to learn to think and do for themselves. Instructor Ian Wade summed it up for me in a Chinese saying by Lao-Tzu: "A leader is good when followed, worst when feared. But, of a great leader, when the task is finished, the people will say, 'We did this ourselves.'"

Expedition school with Rainier Mountaineering Inc. (RMI) was held in the deep snow of winter conditions. Though we were not able to reach the summit of the 14,411-foot mountain, I learned more about winter glacier travel on steep terrain, and at the end of the seminar, found myself hired on as a guide for the summer season beginning when I returned from my planned Mt. McKinley expedition in June.

Lou Whitaker, brother of Jim Whitaker, the first American to summit Mt. Everest, was a powerful climber of great strength and Owner/Guide for RMI. Tall and forceful, Lou plowed his way through the deep snow on our return from Camp Muir, looking back often to see if I was still with him. Years of competitive college cross-country running, track, and cross-country skiing had trained endurance strength into me, and I was able to keep up as Lou post-holed his way down the mountain. Though there were only a few mountain guide services in the United States, I suddenly found myself with a job.

Denali, then more popularly called Mt. McKinley, was next. I knew I needed high altitude experience and signed on with Ray Genet, of Genet Expeditions, to climb the mountain as a client. In June of 1974, I found myself in the village of Talkeetna for the first time, walking up and down the one paved street in my large, vapor-barrier, white rubber bunny boots, also known as Mickey Mouse boots, and crampons, grinding the sharpness out of the crampon teeth. Too sharp crampons were dangerous, for occasionally a crampon would snag a leg while climbing and could cause potential injury.

The weather was excellent as legendary glacier pilot Don Sheldon flew three of us onto the mountain in his silver Cessna 180. The Alaska Range, especially 20,320-foot Denali, 17,400-foot Mt. Foraker, and 14,570-foot Mt. Hunter, was awe inspiring. Standing between these three giant glacial peaks on the Southeast Fork of the Kahiltna Glacier and having to bend my neck back to look up to the summits was a surreal and overwhelming experience. The awesome quiet, punctuated by the occasional roar of an avalanche or the drone of a ski-plane, soaked into my soul. I had never felt peace like this.

Magnificent desolation was one way to describe it. On this trip in 1974, our group was very nearly alone. We saw only a few people the entire trip up the West Buttress and down the Muldrow Glacier. Radio communication was nonexistent until we reached the upper mountain and found line-of-sight to Anchorage. We were alone. Rescue, if needed, was up to us. We could count on no one else. The National Park Service had no presence on the mountain, and there was no Base Camp Manager as there is today. We moved, step by step, camp by camp, up the mountain, feeling very small as we snowshoed over the surface of the 42-mile-long Kahiltna Glacier.

We moved together as roped teams, prepared to catch one another with an ice axe arrest should someone break through a snow bridge hiding a deep crevasse. Crevasses are one of the major considerations in glacier travel. Formed by the uneven flow of the glacial ice over changing terrain, crevasses open and close slowly and are often hidden by snow blowing over the cracks in the ice, forming a weak bridge that can, at times, be difficult to see. The depth of the glacier ice being pulled slowly downhill like a slow moving river is astounding. Glaciologists report a depth of over 3,800 feet of ice flowing at a speed of 3.5 feet per day through the Ruth Gorge, a few miles from the Kahiltna.

Should a person fall into one of these cracks while roped to teammates, they can either self-rescue using specialized rope ascending clamps, or be hauled out by teammates using a z-pulley system, a pulley system rigged from climbing ropes that allows a group to haul its fallen member more easily out of their predicament. Should they fall unroped into a crevasse, the result is always serious and often fatal.

Genet and co-guide Tom Ross were powerhouses, strong and capable. I climbed along with them, watching and learning all I could. Ten days from base camp, we were on the top of the massive mountain, and the highest point on the North American continent. The summit of Denali was out of this world, and I was so affected by the high altitude that I did not recognize my own summit photos after I had them developed!

After a day of rest our smaller traverse team of seven climbers ascended to 18,200-foot Denali Pass while the other members of the expedition descended the standard West Buttress route with Ray Genet. That evening the wind began to rise and as morning came we found ourselves trapped for three days in a cold and bitter storm. Mountain passes funnel and magnify the winds. We were trapped in a high-wind ground blizzard until the storm blew itself out. With seven of us crammed into one four-man tent, we waited out the winds that roared like a freight train passing by, trying to stay occupied and warm. Though it was very cold, it was difficult to stay dry as frost would condense on the tent walls and spray us with each wind-driven flap of the tent wall. The single tent pole would often break through the roof of the tent, leaving us encased in an orange bag of flapping nylon, cracking like a whip in the wind.

Time seems to slow in uncomfortable, drawn-out moments like this. Huddled together, holding the tent down against the winds force by sheer body weight, occasionally sharing a passing thought with the bundled up figure next to you, catching short cat-naps as weariness and boredom overwhelm you only to be rudely awakened as the flapping tent wall would spray you in the face with condensed, frosty ice.

Finally, as the winds calmed, Tom Ross told the group, "time to go," and we began breaking camp, preparing to descend the vast glaciers and steep ridges to the northeast. Worn out from the long and uncomfortable storm, we moved slowly. A couple of hours later, we roped up in teams and began to move.

"Did you hear that?" someone said. "I think I hear a plane." At over 18,000 feet it was rare indeed to see a small plane fly by, but soon pilot Don Sheldon flew low over our team with Ray Genet's bushy, black-bearded face smiling in the side window. Turning and coming around again, Ray leaned out the open door of the Cessna 180 and with his standard pirate, "Aargh," dropped us a care package of food, fuel for our mountain stoves, and ice cream.

I was thoroughly enchanted with Denali. It was truly a new and unearthly world.

We descended to the lower glacier, the realm of warmth and running water, where the smell of rock and tundra overwhelmed us. After three weeks of snow and ice, cold and wind at an altitude that kept even bacteria from living, the sun glittering off small waterfalls and the sounds of birds and falling water entranced us.

Losing the trail from McGonnagal Pass to Wonder Lake, we bushwhacked our way to ford the mile-wide braided streams of the icy McKinley River. I watched as guide Tom Ross tried a channel of the swift, glacial water only to find it too deep to cross. Slowly and carefully slipping out of one of his heavy pack's shoulder straps, he flipped the pack beneath him and used it as a raft, kicking his way to the shore. Hiking the last deep tundra and forest miles of the route to the park road, we were picked up by the last bus out of Denali National Park that evening. As it was too late to catch the train to Anchorage, the rangers allowed us to spend the night in Eilson Visitor Center.

Sitting down is a great luxury. After having nowhere to comfortably sit for the entire expedition, I fell asleep sitting on the toilet in the warm, dry building.

CHANGE IN THE PLAN

Poor weather on Mt. Rainier kept me from working with RMI that season, but a few months later, Ray Genet would find me in Anchorage and offer me a job as a guide for Genet Expeditions. In a few short months the nearly impossible dream had come to be. I found myself with two job offers doing what I loved the most, climbing and guiding others on the high mountains.

It would be a busy summer. Guiding three expeditions on Denali in one season would be a challenge. The first, beginning in May would include a three language group that nearly drove me crazy. The German Alpine Club members were tough and boisterous, full of life and German energy. The lead guide, Frenchman Michel Floret, spoke a little English but no German. I spoke neither German nor French. It was awkward to say the least.

"Does it seem to you the Deutsche Alpine Ventures group is traveling a bit heavy?" I asked Michel a few days out from base camp. Speaking to one of the team who spoke some English, I asked, "Would you please have your team members bring out any extra personal gear that might not be necessary?" One by one, the Germans came with smoked hams, bottles of Cognac, and various other luxuries. We feasted that night, decreasing the load considerably.

Weather was a problem. There were very few clear days, none in fact, on the entire trip to the summit. Travel was restricted to a crawling pace as we probed for the trail under the deep, new snow using a ski pole. Trail wands, bamboo garden stakes with survey tape tied to the top, would be left behind every 150 feet or so to mark the trail. As we stood on the cloud-wrapped summit of Denali after a longer than normal ascent time, it occurred to me that, except for the fact we had no more up to climb, the group would not have known what mountain, or summit, they had attained. The only other evidence they had was a pole stuck in the snow and the fact that one of the climbers threw up due to the altitude.

As we descended the glacier to base camp, the skies cleared and the climbers saw where they had been for the first time. We waited in continuing snow and cloud at base camp for six days to have weather good enough for the flight out to Talkeetna. Our primary diversion was snowshoeing up and down the glacier runway, packing the snow for the ski planes to land on without getting stuck.

Not very heavy to begin with, I lost thirty pounds on that expedition. As expedition climbers often do, craving certain foods, I headed for the nearest restaurant and ate three breakfasts washed down with lots of milk. If I was going to last for all three trips, I would need to do something about my weight.

The something I chose to do when I returned to the mountain, three days after the end of the first climb, was to eat and eat until I couldn't possibly eat any more, and then eat one more spoonful. It worked. I gained five pounds on the next expedition, a traverse of the mountain, climbing the West Buttress route and descending the Muldrow Glacier into Denali National Park.

Some trips have clients who are amazing and great to be with. This was one of those. I remember one of the clients, Jack Heffernan, being so moved that he wept along the entire summit ridge. If every trip were like this, full of appreciative and excited people, I would be happy to do it for free the rest of my life.

Riding the Alaska Railroad back to Anchorage after the descent of the Harper and Muldrow Glaciers, we stuffed ourselves with the prepared food on the train, enjoying the free ride as the rich, green, Alaskan summertime scenery flew by. The second trip was done; the third expedition would begin in three short days, and would be an entirely different experience.

Genet called us in Anchorage and said we needed to get back up to Talkeetna as soon as possible. The weather was taking a turn for the worse and it looked as if we were in for an early fall, with increasing clouds and snow. We arrived on the mountain with the largest group I had ever seen—twenty-five climbers and two guides.

Tom Ross and I, assisted by my girlfriend, Roni, would shepherd this large mass of people to the summit, we hoped. National Park Service regulations were not yet well formed and groups such as this one would often be so large as to be overwhelming. Regulations would be enacted later that would ensure a smaller guide-to-client ratio. Plodding up the glacier with our ponderous packs in late July, on the unwieldy 10-inch-by-56-inch wooden trail snowshoes was a chore in the deep, ever increasing snow.

As we traveled, we got to know the group members; some were climbers, but most of them were inexperienced. Tom and I were concerned about the safety of the group should there be an emergency.

The snow continued to fall. Slowly breaking trail below 13,200-foot Windy Corner through waist-deep snow, we were shocked by the sudden and frequent settling of the snowy slope around us with a loud "whump."

"Did you feel that?" I would exclaim to the climber behind me on the rope. Close to having to quit due to the depth of the snow and the steepness of the slope, I was relieved of the trail breaking burden by a team of Japanese climbers, descending from above.

Avalanches ran from the icy chutes of the West Buttress, making us trek around Windy Corner quickly and nervously, as the wind blast of the powdery avalanches crossed the trail ahead.

Arriving at the 14,200-foot plateau, often used as a type of advanced base camp, we found well over eight feet of snow had fallen in the last two weeks. A few days before, we had left a tall, single-pole expedition tent with food and fuel pitched here. "Do you see any sign of it?" Tom asked. There was no evidence of the tent anywhere in the large, snow covered bowl. With a deep concern, he said, "We need that food and fuel, we have to find it." We probed for the buried tent for an hour, snowshoeing back and forth over the most likely area. Stepping on the top of the seven foot center pole, I yelled, "I found it!" The seven foot tall tent was entirely buried under the newly fallen snow.

We were confronted with a serious problem. We now had the food and fuel to continue the climb, but the deep snow around us was highly unstable. To gain the safety of the ridge of the West Buttress at 16,000 feet would require 2,000 feet of climbing over steep, avalanche-prone slopes. To further complicate our dilemma, a medium sized portion of the slope above the camp had already slid and covered about one-third of the large basin we were camped in with avalanche debris. We needed to make sure we could travel the slopes above safely or we might have twenty-five dead clients. "Tom," I said to my co-leader, "let's climb up to that fracture line tonight and check it out."

That night, Tom and I, along with two of the clients, climbed the already avalanched slope to the north to do a fracture line profile. We wanted to see what had caused the slide. If we could compare that information with another pit dug at the base of the fixed lines we might determine whether it was safe to climb up the steep snow of the route to the ridge above.

We found a deep layer of fragile ice crystals at the base of the avalanched slab. This bed surface, or sliding surface, was unable to hold the weight of all the new snow and had failed, causing gravity to take over and pull the slab of new snow down the hill.

Facing the risk of triggering another avalanche, we climbed to the base of the ropes fixed in the snow and ice from 15,000 feet to 16,000 feet. We dug deep into the steep new snow. "Look at this," I said as we came to the same very fragile weak layer of ice crystals. "This is very unstable." Concerned for the lives of our clients, Tom and I were thinking the same thought, *This could stop us.*

As we discussed what this might mean for our expedition, one of the clients accompanying us piped up and said, "You know how important this is for me. There might be an extra hundred in it for you if you get me to the top." He was offering us a bribe.

Saying we would go back to camp and consider what to do, we descended and tried to catch some sleep. As I rested in my sleeping bag, half-awake, half-asleep, a vivid picture suddenly came to mind. In a dream-like way I experienced our camp being overwhelmed by a large wall of moving snow. This was certainly different for me. At the same moment, Roni and Tom were having difficulty sleeping, too.

Feeling a deep foreboding, Roni asked, "Are you awake?" As we spoke it became clear we were experiencing the same feelings.

"I think we need to move camp as soon as we can." I replied.

"Hey guys, what's up?" Tom whispered from the tent next to us after hearing us talking.

"Meet us in the igloo," I answered.

We met together in an igloo nearby to discuss the situation. We felt a very strong inclination to leave the 14,200-foot bowl early in the morning, before the sun hit the slopes above.

Turning the clients away from their summit hopes was tough, but we had no desire to see them die trying to fulfill their dreams. This was complicated by the fact that only a few of them had enough experience to appreciate the dangerous situation. We descended to the base camp landing strip that day and waited for a plane to come in so we might communicate with Ray Genet about what had happened.

The next plane in had Ray as a passenger. He would often fly in after an expedition had started and catch up with the group near the summit. He was surprised to find us there and immediately began to talk with the clients about what had happened rather than with Tom and me, his guides. The clients were upset. Ray tried to reorganize and take the group back up the mountain. We felt no peace about doing this.

Finally, realizing my dream job was at stake, I agreed to go with Ray to a place farther up the route and be a safety valve, a backup place to leave clients who might not handle the altitude. As I roped up to begin the trip back up the glacier, I felt a great anxiety. I could not, in good conscience, take the clients back up, even in a support role.

The internal pressures continued to build until I could no longer ignore them. I unroped, took off my climbing harness, and, knowing I would likely lose my treasured job as a guide, said to Ray, "I'm sorry, I can't do this." The trip was over as, I felt, was my dream of working on the mountain. I was crushed.

Only a few of the clients were willing to return to the higher mountain, and to go back up without the support of his guides was not practical or safe. Offering the clients an expedition at another time, Genet cancelled the climb and the group began to fly back to Talkeetna and home, wherever that might be.

On the flight out a few hours later, I was impressed by the sheer numbers of beautiful peaks in the Alaska Range as we traveled over the forty-two-mile-long Kahiltna Glacier. A thought crossed my mind, as I looked out the window of the plane, that there were plenty of mountains to go around, and that I might be able to start my own guide service. My dream might not be dead.

Later that fall, I spoke with Dan Kuehn, the Superintendent of Denali National Park, and found I was able to be a permit holder for my own guide service. I teamed up with a great partner and climber, Gary Bocarde, who had been operating a climbing school and guide service named Mountain Trip. The dream was unfolding into a new version as we planned our first expedition together—the 1976 Denali Hang Glider Expedition.

HANG GLIDING
MT. MCKINLEY

The radio telephone call didn't go as we hoped. Climbing above our camp at 14,200 feet on Denali's West Buttress to get line-of-sight communication with Radio Anchorage was a nuisance, but it was all the communication we had with the outside world below. In this instance, the need to communicate was a matter of life and death.

A team of climbers had stumbled into camp that morning with news. "Two Austrian Mountain Troop climbers are down with altitude sickness. One is coma-tose. Their team is evacuating them from 17,000 feet. They need help." On this expedition, rescuing others would become an old story.

On a high altitude mountain, the pace you travel is of critical importance. The human body can adjust to changes in elevation, but it does so slowly. High altitude medical studies suggest an average change of one thousand feet per day is the best schedule for the body to acclimate. Bad weather can be a helpful factor, pinning groups in their camp while a storm blows through, forcing them to rest and giving their bodies time to adjust. This year, the weather was excellent, blue sky day after blue sky day. The Austrians, skilled and strong climbers, had traveled too quickly. At 17,200 feet one was nearly comatose and unable to descend and another desperately ill but able to descend on his own power.

The climbers at the high camp evacuated them to the top of the fixed lines at 16,000 feet. There, we would meet them and lower them to our camp 2,000 feet below. Helicopter rescue was not available at the higher altitude and considered very risky above 14,000 feet. Gary Bocarde, my guide service partner, and I led a team from our camp to meet the rescuers.

Lowering the disabled Austrian, Matthias Knilling, down the steep ice of the Headwall went smoothly and quickly. Anchoring to the ice by the fixed ropes and ice axes pounded into the icy surface, I lowered them both on a rope as Gary guided their descent. The other Austrian teammate, Paul Seitz, stumbled in a sick fog down the trail. We were back in camp with the incapacitated men in a few hours. They suffered from High Altitude Pulmonary Edema, a condition that could cause them to slowly drown in their own fluids as their lungs filled with liquid. Both were also hypothermic, their body temperatures dangerously below normal. There was only so much we could do.

Warming them was something we could do. As we approached our camp, I called out, "We need to get these guys warmed up, now." Warm bodies lumped together under a pile of sleeping bags began to reverse the hypothermia, but they were still in critical condition and needed a helicopter evacuation if we could arrange one. That was the purpose for the radio telephone call to the National Park Service headquarters. We were told they would try, but they could not guarantee anything.

Late in the evening we heard the "whup, whup, whup" of the rotor blades as the Bell 205 Huey helicopter chopped through the thin, cold air toward our camp. "Get them ready!" I shouted to our teammates in the tent. "This is going to happen fast!"

Not being sure what the altitude would do with his turbine engine and not wanting to stall, the pilot kept the throttle on as he lightly rested the skids of his machine on the soft snow. Windblast temperatures were far below zero as we hauled the climbers through the swirling snow and dumped them, unceremoniously, into the helicopter. It all happened in about thirty seconds, and they were on their way to Anchorage and Providence Hospital, where they would recover as their bodies came back into balance in the lower altitudes. We were exhausted and spent the next day resting before we returned to the ascent of the mountain.

There were more rescues, but the most difficult work we had to do was carrying four twenty-foot-long, seventy-pound hang gliders to the summit of Mt. McKinley.

In 1976, Mt. McKinley was just beginning to become popularly known by its Tanaina Indian name, Denali, meaning *the Great One.* The highest mountain in North America at an altitude of 20,320 feet was a magnet for our attention as we flew our hang gliders in the mountains throughout south central Alaska. We would often climb area peaks with the kites on our shoulders to get to the next highest spot and the longest possible flight. As a young high school student, my soon-to-be wife Roni had planned an expedition carrying hang gliders to the summit of Mt. McKinley as a project in a Wilderness Travel class. It was the dream of every hang glider pilot in the area.

During the expedition season of the year before, Ray Genet had mentioned that pilot Buddy Woods had a helicopter he thought could land on Denali's summit. We thought about an attempt to land me on the top of the mountain with my hang glider, but weather and other circumstances got in the way.

Having climbed the mountain as a guide provided a great opportunity to connect the dots and organize an attempt to get a hang glider to the top of the continent. We had four pilots and an alternate that were willing to give it a try.

Climbing the mountain was hard enough without having to carry the added, cumbersome weight of a disassembled kite. Approximately half of those who try to climb the mountain will fail due to altitude problems, weather conditions, or group dynamics. Possibly, we were biting off more than we could chew.

From observation, I felt that about one-quarter of the days in a climbing season were flyable days, days with little or no wind and clear skies. We felt we would have to establish a high camp at 19,500-foot Archdeacons Tower, a prominent feature 800 feet below the summit, and wait for flyable weather. This would be a lot of work.

By the time we got to base camp on the Southeast Fork of the Kahiltna Glacier, we would have not only the four hang gliders, but also a full ton of food, fuel, and equipment to supply our expedition team of twelve climbers for over thirty days.

Arriving in Talkeetna, the jumping off spot for most McKinley expeditions, we waited four days for weather that would allow our flight to the glacier with our pilot, Cliff Hudson. The kites, too long and unwieldy to be flown in by fixed-wing aircraft, would be ferried in by helicopter.

Being at 7,000 feet on the Kahiltna Glacier landing strip between 14,500-foot Mt. Hunter, 17,300-foot Mt. Foraker, and 20,320-foot Mt. McKinley, is an awesome experience. Authors, trying to express its magnificence, have called Denali and the Alaska Range the throne room of the mountain gods. Astounded by the views, the helicopter pilot shut down his machine and got out on the glacier to take photos. In preparation for the trip out, the pilot returned to his helicopter to spin up the turbine engine. It wouldn't start. The fuel injection system was not calibrated for the 7,000-foot-plus altitude at base camp. The pilot hitched a ride back into town on an empty plane, leaving the helicopter where it landed. Two days later, another helicopter would land with a pilot and a mechanic, who would make a small adjustment and reset the engine to operate properly. As they flew off down the glacier, we hoped we might adjust to the altitude as easily.

Commonly, an expedition will carry enough food, fuel, and equipment to require them to double carry or shuttle their camp and equipment up the mountain in two stages. It's common to have 100 lbs. of gear per person. We had so much equipment that it took three trips back and forth to move gear to the next camp. We would work all through the daylight hours, moving our equipment, finally having our evening meal after the sun dropped behind a mountain ridge and the temperatures plummeted to -20 degrees Fahrenheit. One of the greatest challenges we faced was how to eat our food before it froze in the cup!

Our team of twelve people consisted of four hang glider pilots, Bob Burns, Kent Hudson, Mason (Butch) Wade, and Ed Kvalvik; and an alternate pilot, Gene Maakestad. Cinematographer Roger Derryberry, photojournalist Dennis Cowals and his assistant and load-carrying Sherpa, Bruce Hickok, documented the expedition. Two guides, Gary Bocarde and myself and two girlfriend-assistants, Virginia Davey and Roni Bergstedt managed and directed the climbing. A diverse but focused crew, we had only one goal in mind, getting our hang gliders to the summit and safely getting everyone home again.

Day followed day as we slowly ascended the long West Buttress route to the upper mountain. The route itself was only sixteen miles long, but traveling it three full times, together with the descent would add up to over one hundred miles of climbing. Skiing with heavy packs and sleds or hauling the kites filled our days. Temperatures continued to vary widely as the sun would bake us during the day only to forsake us and leave us to the sub-zero temperatures of the night. Camp followed camp as we slowly gained altitude, climbing higher during the day and coming back to lower altitudes to sleep to help with altitude acclimatization. We were getting ready for the upper mountain—colder, steeper, and higher.

Traversing Windy Corner, a narrow glacial shelf skirting the rock of the West Buttress at 13,200 feet, is always breath-taking in its scale. Thousands of feet of exposure open your eyes to the vast sea of ice, snow, and rock of the Kahiltna Glacier and the surrounding peaks. Just above the traverse, where the massive bowl of the 14,200-foot Basin opens up, our roped teams paused for a rest. Roni was at the end of the rope in front. I was at the head of the rope behind her, so she and I could enjoy a moment together, and somewhat alone, as we snacked on our high-calorie goodies.

Roni and I had talked about marriage in the past, and somehow we just didn't get it. We were committed to each other but didn't see the significance of a piece of paper in solidifying our relationship. At that moment, however, something made sense to me. After helping her on with her heavy pack, and just as her rope team began to move, I asked her, "How would you like to be my wife?"

With the rope to her teammates coming tight, she gave me an amazed look and sputtered, "Yes."

Just then the rope tightened and jerked her back to the task at hand. We celebrated that evening at the advanced base camp at 14,200 feet and were married later that summer.

Denali is an arctic mountain, with the last few thousand feet reaching into the upper half of the Earth's atmosphere. Temperatures can easily reach -40 degrees Fahrenheit even in the summer months. Storms can pin climbers down for days. Would we have the weather we needed to pull this off? Did we have the strength to place a camp higher on the mountain than ever before? Could we get these crazy kites through the twisting, mixed rock and ice of the ridge above?

The terrain above the advanced base camp at 14,200 feet is steep and icy, too steep to carry the kites by hand or by pulling them with one end attached to a pack. The icy slope of the headwall from 15,000 to 16,000 feet would force us to haul the kites up with ropes, one by one, foot by foot, to the ridge above, using a modified Yosemite Hauling System, a pulley system used by high-angle rock climbers to haul their equipment up the cliffs they climb. What would normally take hours would take days. The great weather held.

Walking the tightrope of the ridge of the West Buttress presented a unique dilemma. First, it was a series of steep mixed ice-and-rock steps that we weren't sure how the kites would fit through. Second, to travel the last half-mile of the ridge, with 2,000 feet of exposure on one side and 3,000 feet on the other, we would need a near windless day to move the kites. If you've ever tried to carry a piece of plywood in a strong wind, you will understand why. The kites had a strong tendency to weathervane and knock the carriers off balance. This was no place to fall. Again, at just the right time, we had a perfect, windless day.

The individual pilots, wanting to take full responsibility for both the kites and everyone's safety, carried the hang gliders over the very narrow ridge. At times, the nose of the kites would be hanging free over a 3,000-foot drop as the pilot tip-toed his way over a section of the ridge only six inches wide at 17,000 feet.

While at the standard high camp, we became involved in yet another rescue. A team from Colorado was keeping pace with us. To help himself sleep, one of the climbers had taken a strong sedative, phenobarbitol. He couldn't wake up the next morning. Alerting the National Park Service to the situation, they requested an Air Force C-130 Hercules from Elmendorf Air Force Base in Anchorage, to drop medical oxygen by parachute into the high basin. The crosswinds were blowing strongly from east to west as the giant, four-prop cargo plane "crabbed" sideways in the wind as the pilots tried to maneuver above the tents of our camp. From their perspective, trying to drop the heavy oxygen bottles onto our small plateau was like trying to hit a postage stamp.

Concerned that an oxygen bottle might land on a tent and crush those inside, I directed, "Everyone get to the rocks on the ridgeline."

They dropped, and missed, the dropped item falling seven thousand feet to the glacier below. They dropped again and barely succeeded. "Grab it!" we yelled to one another as we scrambled on the ice-covered rocks to grab hold of the parachute lines as the 100-pound oxygen bottle slipped off the ridge. A regulator, one of two dropped, made it down, and we were in business. As the sick climber stabilized by breathing the oxygen, we bombarded the Rescue Gully, a straight drop from 17,200 to

14,200 feet, with rocks to test the stability of the snow, and planned the rescue.

In the light of morning a few hours later, Gary told the teams in camp, "Tie all of the climbing ropes together, end to end."

"I'll drop down the gully before you and set up a belay station with our five hundred feet of fixed line," I told guide Michael Covington. "You guide the patient down the gully as the other teams lower you and I will be waiting below."

Together, we lowered the comatose patient 1,500 feet in just a few minutes.

The patient, Rich Riefenberg, strapped up in a sleeping bag and unable to see out, began to awaken. He began to laugh in a high altitude giddiness, seemingly having fun on his ride. His laughter likely would have stopped if he could have seen the situation he was in. "Sshh," Covington said to us as he guided the patient across a traversing section. "The snow is unstable; be very quiet in your movements." Carefully, we descended the last 1,500 feet without major incident and had Rich in a helicopter and on his way out in a few hours.

The summit was within reach, but the task of transporting the heavy kites and enough gear to survive to our planned highest camp at 19,500 feet was overwhelming.

Late in the night, four of us stood at 19,500 foot Archdeacon's Tower, the site of our proposed high camp, trying to melt ice for drinking water. "Where are the others?" Roni asked as we looked back over the trail below. No one else was in sight.

As we prepared to set up camp, Bob Burns unpacked his small, hand-held radio transceiver. Walking away from the noise of our mountain stoves, he made a HAM radio call to Elmendorf Air Force Base and their weather forecasters. Rushing back from the rim of the plateau where he had made his call he breathlessly said, "They're predicting that the Jet Stream is going to dip down to this altitude within a few hours. They say winds over one hundred miles-per-hour are likely!"

As we scanned the western horizon at 1:00 AM in the Alaskan spring twilight we could see a band of clouds with an upcurling leading edge approaching quickly. We needed to get to a lower, safer camp—and fast!

After the long, difficult day getting to our proposed high camp, the descent in the low light of the early morning was surreal. We found the other members of our team, exhausted and resting in a tent 500 feet below and passed on the bad news that we needed to reverse the days work. A few hours later we arrived at the 17,200-foot camp just after sunrise.

After a few hours rest, we had some difficult decisions to make. It was obvious we did not have the strength to move a camp to the higher altitude. At the same time, Bruce Hickok came to me and reported, "Jim, I've hit an altitude limit. I can't keep any food down." This was extremely dangerous, as dehydration would set in immediately, leading to progressive weakness and susceptibility to frostbite. He had to go down before he would need to be rescued.

We had been on the mountain for thirty days. Gary and I, as guides for Mountain Trip, had another group of clients arriving in a week for another climb of the mountain. One of us needed to go with Bruce to prepare for the next expedition. I had been injured on one of our practice climbs for the expedition earlier in the season and had only been out of a full-length leg cast for one month before the climb. My body, especially my right ankle, was failing. It was decided Roni and I would accompany Bruce, who got stronger and stronger as we descended. We skied into base camp after an all night descent of the route and were flown out by Doug Geeting, then a brand new pilot working for Cliff Hudson. We were back in Anchorage twenty-four hours after leaving the high camp. The jet stream never dropped. The winds never materialized. The weather remained perfect.

That next night, the hang gliders and the pilots reached the summit of Denali at 20,320 feet. The weather was optimal with temperatures slightly below zero and no wind. It had been planned the flyers would use skis to launch. Due to the altitude, no one knew precisely how much air-speed was required for sufficient lift to fly. Since the pilots would be wearing full expedition gear and foam flight harnesses to withstand the cold, and be carrying emergency rations should they be forced to land somewhere other than base camp, as well as carrying the weight of the hang gliders, to launch on foot in calm conditions was considered nearly impossible.

But the pilots decided it wasn't true hang gliding to ski launch off the 10,000-foot South Face. They chose to attempt to foot launch. A runway was stomped out down a small rib extending from the summit down the immense South Face. After thirty days on the mountain, it was time to fly! Butch Wade yelled, "We're going to do this!" as Bob Burns prepared for takeoff.

Filmmaker Mike Hoover flew near the summit in a Bell Jet Ranger helicopter. As Mike waited with his movie camera in the rotor-blasted, frigid air on a perch at the open door of the helicopter, Bob ran and leapt from the summit.

Bob was carrying his HAM radio strapped to his wrist and was transmitting to an AM Radio station in Anchorage that was in turn doing a live broadcast of his flight. I listened on the radio at Kleen Fun Kites in Anchorage as he gasped in high altitude hypoxia, "I almost died." Bob had hit his wing tip on the rib of snow as he launched and put the kite into a flat spin, like flipping a card, as he left the summit. Looking behind him, he saw no other kites in the air. Frightened and disoriented, he climbed into the triangular control bar and put the hang glider into a shallow dive, dropping over 13,300 feet to base camp in approximately thirty minutes. Butch Wade had followed him off the summit but was not so fortunate.

As Butch ran, and dropped into his flight harness, the triangular control bar of the kite hit the snow. He began to somersault down the steep rock and ice of the South Face. 500 feet below, the broken pieces of his kite jammed into the snow, stopping him from the certain death of a 10,000-foot fall to the glacier below. The need to rescue Butch was now the highest priority—Kent and Ed would have to wait for their turn to fly.

Strangely enough, right at this time two women arrived at the summit to report that their teammates in an all-woman climb of the South Buttress were just below the summit and in need of rescue. Our team was overwhelmed and under-equipped to deal with it all.

In an odd, three-way conference call between Roger Derryberry on the summit of the mountain, Chief Ranger Gary Brown at Denali National Park Headquarters, and myself at Kleen Fun Kites in Anchorage, it was decided the Air Force would again drop rescue equipment to the team at the Football Field, a large flat area below the summit. No other options seemed possible. The equipment, an Akio Sled and rope, was dropped from a C-130 and never seen again. We needed alternatives A.S.A.P.!

Buddy Woods, a very competent fixed-wing and helicopter pilot, had a helicopter he believed could land on the summit. In a record-breaking attempt, jury-rigging the balance and trim on his machine by hanging a five-gallon fuel tank from the nose of the helicopter, he landed acclimatized mountain guide Ray Genet on the very summit. Ray quickly descended to the sick team on the South Buttress and prepared them to be air lifted off the mountain.

While all this was happening, crashed pilot Butch Wade, miraculously uninjured except for massive bruising, had dug out a small platform and was resting on the steep snow of the South Face, waiting for his teammates to climb down to him and get him out of there. Using broken pieces of his hang glider, he jammed them in the snow below him to keep from slipping off the small ledge he had hollowed out. He waited through the night as teammates kept yelling to him from the summit in an attempt to keep him awake. Traversing under the rock below the summit, Gary Bocarde climbed down to him the next morning and helped Butch as he made the climb back to the summit ridge. He would have to walk off the mountain he had hoped to fly down. He felt robbed, but lucky to be alive.

Everyone was rescued, the weather was still wonderful, and even though the Park Service had decided it was imprudent to make another attempt, Ed Kvalvik and Kent Hudson had hang gliders waiting on the summit and would not be denied their flight.

Climbing to the summit two days later, with photojournalist Dennis Cowals and cinematographer Roger Derryberry, they launched in moderate winds to soar over the summit and enjoy a few hours long flight to base camp at 7,000 feet. They were so high above the ground that it was impossible to determine whether they were moving. It was beyond description. Trusting in their Ultralight Products' Dragonflies to trim and fly properly, they enjoyed their dream descent.

As the rest of the team came off the mountain, they gathered gear and sleds, making the long descent to base camp and a ski-plane flight out to the lowlands. For me, one of the greatest thrills of the whole trip was having everyone home alive and safe.

Later that summer, we would lose two of our team, Pilot Bob Burns and Alternate Pilot Gene Maakestad, to unrelated hang glider accidents. Experimenting with flight and reenacting the development of the airplane was proving costly.

Gary and I would return to the mountain in just a few days to successfully accomplish the Kahiltna-Muldrow, West-East Traverse with clients. Compared to the Hang Glider Expedition it seemed ridiculously easy.

MOUNTAIN TRIP

Co-owning and directing our own mountain guide service offered my partner Gary Bocarde and I many new opportunities for exploration and adventure. Business relationships with other adventure travel agencies opened up, and our mountain guiding became international in scope. We were in at the beginning of the adventure travel boom of the middle '70s and '80s. People were coming from all over the world to climb in Alaska, especially Mt. McKinley.

One such client was Will Davis. Sitting together at the top of a rock climb in the Talkeetna Mountains, enjoying the feeling of relaxation that comes after a climb and the warm colors of a late-evening sun, Will, a project manager on the North Slope for oil company BP-Sohio, began to quote *The Spell of the Yukon* by Robert Service.

> *I wanted the gold, and I sought it;*
> *I scrabbled and mucked like a slave.*
> *Was it famine or scurvy I fought it;*
> *I hurled my youth into a grave.*

I wanted the gold and I got it—
Came out with a fortune last fall, —
Yet somehow life's not what I thought it,
And somehow, the gold isn't all.

Will was looking for something more. He thought he might find it in the mountains and wanted to climb Denali. We trained together and got him ready.

There's an old saying that goes, "What you fear will come upon you." Will was nervous about a few things; the power of the mountain and water-filled crevasses to name a few.

Will traveled well and was an energetic team member. He had a great attitude and was always ready to go. As we climbed the West Buttress route on Denali, followed by a descent down the Muldrow Glacier, Will had more than his share of excitement.

One was being a kite. Climbing up the narrow and exposed ridge above 16,000 feet on a very windy day, Will was hit by a sudden and strong gust of wind that picked him up and threw him. Fortunately, a teammate held his rope, pulling him down and back on to the ridge.

When a person is imprinting a newborn horse one of the techniques to prove to the young horse that you are more powerful than he is, is to pick him up and get his feet off the ground. After flailing for a moment, the foal will accept that you are bigger than he is and that he can trust you. Will definitely got imprinted. He now really knew the mountain was bigger and stronger than he was. He was, pardon the pun, blown away!

Later, after a successful summit, he had to descend through the Lower Icefall of the Muldrow Glacier, well known for its water-filled crevasses. Will had developed a phobia about falling into a water-filled crevasse.

Falling in one of these crevasses is a deadly situation. Any crevasse fall is bad, but to fall into numbing ice water at the bottom (pooled-up melted water from the glacier), is extreme, to say the least. If you are not well prepared and things don't go your way for a quick rescue, you will likely die.

As I mentioned, the Lower Icefall is famous for this. Often, as you snowshoe over a crevasse, you can hear the water running below and hear ice splashing into it as it is knocked loose by the crossing of a snow bridge. Will was more frightened than most.

As we approached the icefall, Will tripped as he stepped over a tiny six inch rivulet of water running on the surface of the glacier. "Help me, Help Me! I'm going in!" he yelled in his southern accent as he fell flat on the glacier surface. The toes of his boots got wet, but he was safe. Chagrined and a bit embarrassed, he traveled through the rest of the icefall without incident, to his great relief.

AVALANCHE ON MT. SILVERTHRONE

My co-guide and good friend, Bruce Hickok, and I watched the Alaska Range go by under a rising, full moon as we rode the bus out to the park entrance from Wonder Lake. The peaks to the east of Denali, Mt. Silverthrone, Mt. Brooks, and Mt. Mather, swept upwards in graceful ridges of ice and snow many thousands of feet and were stunningly beautiful in the bright moonlight. The moment was idyllic and enchanting. As we rode the shuttle bus, following a successful traverse of Denali, we plotted and planned how we would return after the guiding season ended and climb these peaks.

Fast and light was the plan. Minimal food and gear, a small, light tent, and lightweight snowshoes would enable us to travel quickly into the Alaska Range and up our chosen peaks. This is a good concept, but Alaska weather conditions or an unforeseen problem can turn this philosophy into a nightmare fight for survival.

After an easy crossing of the mile-wide braided streams of the McKinley River, we continued on up into the foothills, enjoying the warmth of the sun and late season lack of mosquitoes due to frost. Sunny, blue skies remained for nearly the whole month of August, a rarity to be enjoyed. Being in good shape from the climbing season, we traveled quickly over the bare glacier ice that transitioned to wet snow and then colder, dry snow as we gained elevation. Two days into the climb we arrived at 10,000-foot Silverthrone Col, a large saddle on the ridge between Mt. Silverthrone and the Tri-Pyramid Peaks, and made our high camp. We rested in our small tent and rehydrated, but didn't sleep much, as the summit of 13,219-foot Mt. Silverthrone was just outside our tent door.

As it so often does, the weather began to change as morning came on. The wind picked up, and we could see snow being blown from the summit ridge in a wispy banner as we prepared for the day's climb. We felt an urgency to get up the peak soon, before the weather really closed in.

Moving rapidly on our small, lightweight snowshoes, we came to the base of a broad slope leading up to the Northeast Ridge at around 11,500 feet. Looking at the slope, we picked an icy section that would provide the best access to the ridge above. Taking off our snowshoes, we took the time to tie them together and leave them at the foot of the climb and use our crampons to ascend. I led on up the slope, kicking steps in the firm, iced over snow.

Step by crunching step we were rapidly gaining altitude when I stepped up and onto a different kind of snow. Suddenly I became disoriented. Everything was moving around me. I was losing my balance and said to myself, *I'm falling*! Too late to shout a warning, I realized a dreadful fear, a*valanche*! The whole slope above was sliding down on top of us.

Being caught in an avalanche is much like being caught in a large ocean wave; the power of the moving snow is incomprehensible. Your body is pushed and pulled in every imaginable way. You are totally, frighteningly out of control. I was buried upside down repeatedly, only to be ripped out and reburied. For one brief moment, when I had come to the surface of the sliding snow, I saw Bruce in the air above me. He had unexpectedly experienced the climbing rope piling up at his feet as the avalanche approached. When the wall of snow hit him at the knees it literally blew him up and into the air. As he twisted and turned in the avalanche the climbing rope wrapped around him like a cocoon, totally immobilizing him. After running a few hundred feet, the moving snow slowed and finally stopped. I was buried, but able to break out of the avalanche debris and find Bruce, totally tied up by the climbing rope, but on the surface.

"Bruce, you okay?" I yelled as I waded through the debris to release him.

"What happened; where did that come from?" he said.

"I don't know," was all I could say. We couldn't figure it out. The weather had been perfect for days with little or no snow for weeks. *Where did that come from?* was a question that would haunt me for years.

We had another problem. Our snowshoes were buried and missing. It would be a long and difficult return trip down the glacier without them. Soft, deep, wet snow and crevasses would be tough to get through on foot. We searched and searched in the rising wind. The avalanche debris was soon blown over by drifting snow. We set up a search grid, desperately looking for the snowshoes, aware that the slope above was reloading with wind-driven snow and could slide again.

Hours later we had given up and resigned ourselves to the long slog out when I spotted a tiny glint of reflected sunlight in the snow. A harness buckle! We found them. Since we had taken time to strap them all together when we left them, they were all there. Joyously relieved, we gave up our hopes for the summit and returned to our high camp, thankful to be alive.

We probably should have begun descending immediately, but we were exhausted and hurt. Our bodies ached from the twisting and wrenching in the avalanche, so we chose to rest.

The winds were roaring in a few hours as the storm increased its intensity. Lying in a small two-man

Omnipotent, a tent so small we could not sit up in it, we waited. The storm blew, the snow fell. We rationed our small supply of food, beginning a dangerous process of conserving food while possibly not eating enough to sustain the energy we would need to get out of there when the time came. Our backs ached; we waited. With nothing to read, we talked ourselves out.

We waited.

The storm blew incessantly, with wind gusts rattling and occasionally flattening the mountain tent. We lay in the tent trying to nap but unable to sleep. Two days later, the break in the weather came.

It was time to go.

Thirty-five miles of glacier, wilderness, and river crossings separated us from the road and food. We began the slushy slog down the glacier. All through the day and the next night, we snowshoed and walked over glacier and tundra trails, crossing streams in a fog of weariness and hunger.

Twenty-seven hours later, we emerged at Wonder Lake. There had been little reason to stop along the way except to find a drink of water in the streams we crossed.

Returning over the mile-wide braided stream of the McKinley River was a gripping and emotional exercise. It was always serious, being fast, ice cold, and opaque with the grey-colored grit of glacier-pulverized rock. But, this time, we were coming into it traumatized, weak, and anxious, or at least I was.

My heart was racing and my breathing was shallow as I admitted, "Bruce, I am really scared."

He was confident and encouraged me, saying, "Just stick close. We'll do this as a team."

We waded into the frigid water and began to travel through the multiple icy channels of the river below the Muldrow Glacier. Halfway through, the situation was totally reversed, as I now felt strong and confident, yet Bruce was hyperventilating with anxiety.

With a fearful look on his face, he disclosed, "Man, I'm freaking out here."

"It's okay, we'll make it." I replied.

We helped each other through, locking arms as a team and supporting one another in the strong current until we emerged at last from the final swift, waist-deep channel of the river.

Later that year, I was sent by the Alaska Mountain Rescue Group to the National Forest Service Avalanche School in Reno, Nevada. I had serious questions. Taking time with Knox Williams, one of the school instructors, I explained our nearly deadly experience and asked him, "What happened?" Not having technical data to work with, all he could say was, "I don't know." I knew what had happened to us, but I was unable to see ways the accident could have been foreseen and prevented. The school helped me understand principles, but I was still bothered by how to understand and apply these principles to our individual situation.

The following year, my wife Roni and I led a trip back to Mt. Silverthrone and Mt. Mather. I was apprehensive as we approached the same slope that had very nearly killed Bruce and me the previous year. Digging test pits in the snow, I checked for weakness in the snowpack that could fail and cause an avalanche. I found no glaring problems, and we continued up a narrow ridge to a beautiful summit.

It was years later, as I reviewed the slides of the first trip for an avalanche class I was teaching, that I had an "Ahah!" moment. The icy slope we began to climb had been the bed surface of a freshly run avalanche. The slick ice had formed over the past month's excellent weather. The step up onto a new level was stepping onto the Crown Face or Fracture Line of that same recently run slide. It was likely the avalanche had run just moments before we arrived as the bed surface was clean of any wind-blown snow. The slope above us, slightly lesser in angle, was loaded with a new layer of windblown snow that fractured and slid down on top of us as we disturbed it. We had blindly walked right into a hair-trigger trap and fortunately survived.

It was deeply ironic that my friend, Bruce Hickok, would die years later in an avalanche just as predictable as the one on Mt. Silverthrone. The mountains can be a harsh judge of inattention.

DANCES WITH MOSQUITOES

Alaska's unholy trinity of the wilderness is bugs, bears, and brush. Swarms of blood-sucking insects drive the animals to keep moving constantly. They manage, but we humans need to camp and, possibly, sleep every now and then. From necessity, we developed a method of keeping the fog of mosquitoes from entering our tents.

The entire group would mosey on over a hundred feet or so downwind of the camp, drawing the mosquitoes along with us. The group would then brush every bug from one person, who would then sprint into the wind to the tent. Quickly opening the mosquito net door, he or she would dive in, seal the door and kill every insect that managed to follow.

Allowing a few minutes for the pursuing horde to return to the group, the process would then be repeated until we were all in the tents and safe from the blood-sucking bugs.

Travel through the tundra lowlands to and from the high mountains can involve a few days in the mosquitoes' domain. Whenever we had to travel through these habitats we had to take survival steps. Some people wander the tundra like alien bee-keepers just arrived on the planet wearing full Haz-Mat suits. Covering yourself entirely works, but it sure is hot in those suits on a sunny day.

We mostly used DEET mosquito repellant, preferably the strongest mixture you could buy. We found a drawback side effect to the 100 percent DEET, though. One trip, as we covered the twenty or so miles from the Muldrow Glacier to Wonder Lake at the end of a Denali traverse, we found our hands turning green. It turns out that the DEET chemical, at the 100 percent concentration, was an effective paint remover and was taking the paint from the ski poles we were carrying!

It's difficult to describe the numbers involved in the swarms of what is jokingly called the Alaska State Bird. Let me use an example. Sitting on the banks of Clearwater Creek in camp one evening, I began to randomly clap my hands together. In just a few minutes, I had a small pile of dead mosquitoes, an inch or so thick, lying in front of me.

Another competitive game in bug country is gaining bragging rights over your companions through killing more mosquitoes in one swat than the others. I think the record is around seven, though it could be higher.

Mosquitoes are little heat- and carbon-dioxide-seeking missiles and often end up in your food. I'm not sure they add much food value to a meal, but they sure are tasteless.

ROCK FALL INTERVENTION

The thought of exploring the Chigmit Mountains, a group of glaciated peaks on the western shore of Cook Inlet south of Mt. Spurr, intrigued me. My interest began as I viewed the mostly unexplored mountain range from high on the Southwest Ridge of Mt. Spurr, many years before. Consisting of many peaks of frost-shattered granite, we wondered if the mountain range held some opportunity for new rock climbs. We had to go see.

Mark Moderow, our guide service lawyer, and his girlfriend Margit Bretzki, would join my wife and me on the exploration. Flying in to glacial Chakachamna Lake with Ralph Eldridge in his Cessna 206 on floats, we began the approach. The Chigmits hadn't been visited very often. We were unable to find any information at all in the *American Alpine Journal* or any other resource. As we were climbing to the glacier level, a passing Air Force para-rescue helicopter spotted us, and, for a moment, lowered its wheels in anticipation of a rescue. They probably hadn't seen any climbers in the area, ever. Seeing that we were equipped with packs and not the victims of a plane crash, they waved and flew on.

The mountain ranges of south central Alaska, with one slope facing the dry interior and the other the influence of the North Pacific Ocean, experience quick and powerful weather changes. Our weather now changed.

It rained for days.

We sat in our tent talking and reading books as the storm pelted the tent walls with cold rain. To stave off boredom, we would dress warmly, rope up, and climb small glacier seracs or towers, just for something to do. After a climbing session, wet to the bone, we would pack all our soaked gear into a plastic bag, change to our dry clothes and rest. This kept our sleeping area dry and warm. Putting on that wet, cold gear to go out again later was excruciating. It was hard to imagine anything could be colder as the near freezing, wet clothing touched our skin.

Finally, after a few days, the weather began to break. We were able to explore the beautiful high basins surrounded by 7,000-foot peaks, climbing small spires of rock here and there. The rock was a badly weathered, decomposing granite offering little hope of longer, high angle climbs.

On one particular loop through the high country, we had a steep snow and ice slope that needed to be crossed to return to our camp. It evidenced signs of previous rock fall, the tell tale scars and black debris from rock falling down the slope from the cliffs above. In the wet days before, we had heard no sound of falling rock and determined it was reasonably safe to cross if we crossed late in the day when the sun was off the slope. At dusk, after a long and enjoyable day, we began to cross the slope as a four-person rope team. Just in the middle of the traverse, roped together and stretched out the full 150-foot rope length, carrying packs, wearing crampons and holding ice axes, the worst happened.

We heard a thunderous roar above us as a massive rock fall tumbled from the cliffs above. It was so dark that we could not see the cliffs; we could only hear the rock as it crashed onto the slope we were crossing and headed right for us. The thoughts, *Should we try to run, roped together with all our glacier gear? What if we fell? Which way should we run?* swiftly flew through our minds, when something amazing and powerful occurred. Without ever thinking the thought, the words, "Stand and watch," came from my mouth in a deep, commanding tone. We stood and faced the oncoming storm of large rocks.

The rocks were big, chest-sized and coming fast. As they hit the slope and bounced, ice and snow would spray in all directions. The sound of the rocks, humming in their high-speed flight, filled our ears. We were right in the middle of the bowling alley, stretched out over 150 feet of climbing rope.

The rocks all passed between us. Only once would we need to react. "Move toward me," Mark spoke to Margit, allowing the rope, stretched between them, to drop so a large rock would pass over without snagging it and pulling us down the slope. Shocked, we left the slope at an easy walk and returned to camp. I don't believe we ever talked about that supernatural experience after that. Sometimes, there just aren't any adequate words.

CLIMBING ICE

Much of Alaska is covered with ice. The glaciers and the waterfalls that would freeze into solid pillars in the cold winter months provide ice climbers with a multitude of opportunities for exciting climbing. In December of 1977, technical ice climbing, the climbing of the long, vertical waterfalls, was in its infancy. The tools we used, the ice axes, crampons, and ice hammers, compared to those available today, were the equivalent of fishing with a club.

Late that month, a friend, John Dillman, suggested, "Hey Jim, what do you think about going down to Valdez and trying Bridalveil Falls?" Intrigued with the possibilities, we drove the two hundred and fifty miles to Keystone Canyon, near Valdez. Up to this time Bridalveil and its longer companion, Keystone Green Steps, had only been climbed in multi-day efforts. We wanted to see if it could be done in one push.

John and I packed up the Mountain Trip company van and headed off, stopping in the canyon to climb one of the smaller waterfalls, Horsetail Falls, across the canyon from Bridalveil. Water was running over the ice, making the ice soft and plastic, easy to climb but wet and cold. After the short, afternoon climb we retired to Valdez for dinner and to visit the local Laundromat, where we dried our climbing ropes and gear in their large dryers. We received a number of strange looks from the housewives doing their wash as we worked on our bizarre-looking, ice-climbing tools and pulled piles of spaghetti-like climbing rope out of the dryers.

Late that night and back in Keystone Canyon, sleeping in the van, John and I were jarred from sleep by the flashing lights and megaphone of a State Trooper, inquiring why we were there. We explained our plans to the Trooper, who likely thought we were certifiably crazy, and he left us to return to our fitful sleep.

Early the next day, John and I snowshoed over the Lowe River to the base of 600-foot Bridalveil Falls. Beginning at first light, the first few hundred feet of the climb went smoothly. The ice was wet and accepting the placements of the ice hammers and crampons without fracturing, making the climbing fairly secure and comfortable. We were at the base of the crux vertical ice pillar, 450 feet off the river below, by mid-day and had lunch, looking over the 70-foot-tall vertical pillar above for the best route.

As I led off and began to climb the pillar, it became immediately obvious the quality of the ice had changed. It was cold, dry, and brittle. Placing an ice tool or a crampon would result in fracturing, with large chunks of ice coming off, threatening my partner, John, as he watched the rope and belayed below. It was hard to place protection, in the form of ice screws we would put in the ice to hold us in case of a fall. As I would begin to screw one into the ice, it would freeze in place and resist being put in further. I tied off the screws wherever they stopped with nylon sling material, usually just three to four inches into the ice. It was doubtful they would hold a fall. The farther I climbed, the greater the risk of a long fall and hitting the ledge below. I came to the point where an attempt to descend would be far more hazardous than to continue to climb upward, as climbing down vertical ice is far more awkward and difficult.

Forty feet off the belay ledge, and 500 feet from the river below, my front crampon points stripped out of the ice as I was moving an ice hammer. I fell hard onto my one remaining point of contact, my other ice hammer. My hand slipped off the handle, and I was left dangling by my wrist from a one-half inch nylon strap, with my back to the ice and my feet flailing freely below me. I looked down 500 feet as John quietly said, "Stay cool, hold it together," from the ledge below. That wasn't easy, but I knew the pick of my ice hammer was only an inch or so into the ice; I could not move violently or the ice around it could fail, sending me on a certain, long fall. Strange, "eep" sounding noises issued forth as I struggled to regain a secure position.

With my heart beating out of control, too frightened to take a deep breath, I managed to twist around, get hold of the ice hammer handle and place my crampons in the ice again. Quieting my heart and breathing once again, I continued up the vertical ice and finished the crux pitch.

John followed and then led off to the next and final pitch to the top of the frozen waterfall. With our adrenaline flowing, we celebrated in the fast-approaching dusk. The celebration was a short one as we prepared the frozen ropes for the descent. We knew we had to get down quickly or be caught in the night.

As we rappelled down the waterfall into the deep canyon, it became very dark very quickly. There was little direct sunlight just a few days after the shortest day of the year.

In the middle of the descent it became totally dark. We were unable to see the ice clearly and spent time groping like blind men trying to find good placement for the ice-screws we would leave behind as anchors for our rappels. Somehow, we had neglected to bring headlamps, a flashlight, or light of any kind.

Two hundred feet from the bottom of the waterfall, the ropes we had just slid down froze in place, keeping us from pulling them down behind us and setting up the next rappel. We worked and worked trying to free them, and were just about to accept spending the night stuck on the ice, when I saw John tying himself off to a six-foot-long nylon runner. I watched, incredulous as he jumped into the darkness, holding on to the end of the climbing rope using a Jumar ascender. His body weight jerked the rope free from the ice, and we were able to pull the ropes and continue down, arriving on the ice of the river below in the pitch black of night.

Climbers of today often solo climb the waterfalls, enjoying good conditions and the higher technology available, but for us, at that time, the first one-day ascent of Bridalveil Falls was the ultimate challenge.

ACCIDENTAL SUMMIT

I had always desired to climb Denali's North Summit. Not often visited and out of the way, only a few groups since the Pioneer Climb of 1910 have been there. The Pioneer Alaskans who did that audacious climb on a bar bet in Fairbanks had seen Mt. McKinley from the north. It appeared from Fairbanks that the North Peak was the tallest, so it was the main summit as far as they were concerned. Actually beginning the climb in December of 1909, they mushed their freight sleds in from Fairbanks. Miners, mostly from the Kantishna District, just under the massive 14,000-foot tall Wickersham Wall of the North Peak, made up most of the party. The Pioneers, equipped with little climbing gear, homemade ice axes, and crampons, carried a fourteen-foot spruce pole to the northern peak of the mountain and planted it in the highest rocks of the ridge now known as Pioneer Ridge for all to see. The ascent route from the Harper Glacier was a steep, icy couloir now named for them: Sourdough Gully.

Farther up the Harper Glacier, just off Denali Pass at 18,200 feet is a much easier route. Climbing an ascending traverse across a band of black, baked mud rock is a fairly easy ledge of snowy rock. Many people look at it, thinking about climbing both the North and South summits of the mountain. However, after an ascent of the primary South Summit, climbers are usually very tired, wanting to escape the mountain and head home for some good food and a hot shower. It was like that for me for years. Circumstances, storms, illness, and other needs often kept us moving rather than hanging around at that altitude for another day.

This time it was different—accidentally.

In 1977, our Traverse team was an especially strong group and the weather was reasonable. We had climbed to Denali Pass the day before and set up camp. Early in the morning, we headed for the South (Main) Summit and moved quickly, arriving at the summit in a little over three hours in a stiff, cold wind. Air Force para-rescue Sergeant Bob Lapointe, a strong and well-trained climber, suddenly began exhibiting symptoms of High Altitude Cerebral Edema just as he arrived at the summit. He began to lose consciousness and the ability to speak coherently. We needed to get him down as fast as we possibly could.

Putting one strong climber on each side, each one holding onto Bob's pack like a harness, we jogged down the slopes toward our high camp, dragging Bob between us when his strength faded. "Hang in there Bob, stay awake, stay with us," we would say to him to help him keep focused. Taking a moment to rest, Bob sat on the snow as Dr. Jim Sprott and I knelt nearby. As Bob went unconscious and fell backwards onto the snow I yelled, "Jim!" to get the doctor's attention.

Springing to his side, Dr. Sprott forgot he was attached to his ice axe by a nylon webbing leash. The ice axe came flying in behind the doctor, and hit Bob on the side of the head with a loud, "Clang." Fortunately, the axe hit him with its side rather than the sharp point. Bob immediately woke up, saying, "Huh, What?" and stayed awake from then on. Later, I jokingly told the group, "I guess we can call that the pre-cranial thump resuscitation technique."

Bob recovered quickly as we dropped altitude and was back to his old self, but tired, when we arrived back at our high camp in Denali Pass. He wanted to rest there, and as he was doing well, we relaxed.

As the others crawled into the tents, I looked at the ascending traverse to the plateau below the North Peak and decided to wander over and take a look. Vern Tejas, then a client but later an outstanding mountain guide on major mountains all over the world, was standing nearby with his crampons still on.

"Want to go check out the route to the North Peak?" I asked.

"Sure," Vern replied, and we walked off, just planning a little reconnaissance. Since we were not planning to go far, we took no gear, ropes, or packs, just ice axes and crampons.

The rocky traverse was surprisingly easy, and very quickly we arrived at the plateau below the North Peak. As we stood there, realizing the difficulty was behind us we decided to keep going on to the North Peak. It was "just over there." We couldn't pass up the opportunity and, soon after, were standing together on the North Peak, just below the ridgeline. We had climbed both summits of Denali in five and a half hours. By accident!

The view from the North Peak is unique. Wickersham Wall, one of the largest mountain walls in the world, over 14,000 vertical feet, is right below the summit. We really wanted to look down it, but we were concerned there might be a cornice on the ridge. If there was, the cornice, like a fragile wave of ice, could collapse and take us with it. We tried to get a side view of the ridge line but were unable to do so.

Remembering an old photo I had seen, I thought there was no significant cornice. Trying to make us as secure as possible, I dug into a seated position and held on to Vern's crampons as he slithered over to the edge on his stomach. Although it was a most uncommon solution and not at all something to be recommended, it turned out to be quite safe as there was no cornice. Amazed and alone on the top of the North American Continent, we lay on the snow of the summit in an unexpected and wonderful moment, gazing over at the South Peak and down the mass of Wickersham Wall.

TOO FAT TO FLY

Raven, in native Alaskan folk-religion is the trickster, like the Loki character of Norse mythology or the coyote of the American Southwest tribal folk-religion. Highly intelligent, they have been seen as high as 18,000 feet on Denali and can cause serious damage to an expedition's food supply as they attack food caches left on the glacier, sometimes digging six inches or more into the snow to find food. Diligent education and enforcement by the National Park Service have taught most modern-day expeditions to be extra careful. In the past, expeditions have tried everything imaginable to keep the ravens out of their food, even packing their supplies in plastic garbage cans.

We had a cache hit by a raven at 11,000 feet at Kahiltna Pass. He gorged himself to such a degree that he had a terrible time trying to fly away. When we found the cache and the bird's wing imprints on the snow the next day, the tale was obvious.

Trying to take off, but unable to generate enough speed at that altitude to lift his stuffed body into the air, the raven hopped and flapped over fifty feet, leaving distinct imprints of his wings in the snow at each flap before gaining enough speed to fly.

TAYLOR CREEK HEAD PLANT

Backcountry skiing can be difficult. Snow conditions can vary greatly, from light, powdery snow to heavier, wind-driven, drifts. Sudden changes in snow quality can create significant surprises as a skier is thrown forward or back, off balance in the inconsistent snow. Snow conditions deep within the snowpack are also constantly changing in a process called metamorphism. This process and its disastrous and deadly results were what we were here to study.

In March of 1982, Roni and I, with our young daughter Crystal riding in a backpack, skied up into the beautiful Taylor Creek Valley, high on the south side of Turnagain Pass, south of Anchorage, with a group of students and instructors from the Alaska Avalanche School. We were there to visit the site of one of the most deadly avalanche accidents in Alaska history and learn why it happened and how it could have been avoided.

A few years before, a group of five ski tourers climbed into the high valley with the intent of making a loop ski tour up through Taylor Creek and down the Lyon Creek drainage. In good conditions, it would have been a challenging but wonderful trip. In the conditions of the moment, four of them would soon be dead.

On that same day, my wife Roni was guiding a graduation ski tour in the flat of the valley below with students who had completed her Beginning Cross Country Skiing class. The snow was deeply unstable, settling suddenly with loud "whumph" noises and a visible drop in the surface of the snow. Roni came home that afternoon and told me of the frightening instability of the snowpack.

Later that evening, a phone call came from the Alaska State Troopers. The Alaska Rescue Group, of which I was a part, was being called out to search for four missing skiers, who had been caught in an avalanche in the Taylor Creek Valley. We hurried to the site, a two hour drive from Anchorage, realizing that every minute counted if the victims were to survive.

We had only sketchy details. The report came from a man, one of the skiers, who had been with the group when the avalanche hit. They had just been discussing how frightened they were as the snow fractured all around them and began to slide. When the snow came to a stop, his friends were gone and he was alone. Tumbling in the snow slide, he had lost his glasses and could not see well enough to know if the others were buried, injured, or worse. His skis and pack were gone. He yelled but heard nothing. Searching a while, with no results, he began a long and extremely difficult return to the road a few miles away.

He had followed the seemingly easiest path, but that led him into a terrain trap of a deep, narrow, snow-and-rock-walled canyon. Deep in the canyon, unable to see in the approaching night, he floundered through icy running water and deep snow to the road.

Hitchhiking to the nearest telephone, he called the Alaska State Troopers to report the incident. Sending a trooper to verify the situation before rescuers were called, precious hours were lost.

Buried avalanche victims usually don't live very long. If they haven't been killed by traumatic injury, ice begins to melt around their bodies forming an ice mask, especially around their faces, as they exhale warm, moist air. This eventually seals off the air space causing victims to asphyxiate. While some have lived for days, half of all those caught and buried perish within the first thirty minutes, and few live beyond an hour or so. There is no way of knowing whether they are dead or alive until you find the victims, and we felt the urgency to get to the avalanche site and begin the search.

Arriving at the Taylor Creek parking lot, we took stock of who was there and how well they were equipped. We might be on site for a while. The weather was deteriorating, with intermittent clouds blowing in from the south, hiding the full moon. A group of us, from Anchorage and Girdwood, started up, leaving others behind to direct the rescuers who would come later. Our headlamps throwing light into the darkness ahead of us and using climbing skins on the bottoms of our skis, we made good time to the entrance of the upper valley.

Avalanche debris filled the valley. We knew the avalanche that caught the skiers had run farther up the valley toward the pass, but since that time, many other avalanches had run naturally, without a human trigger.

The light from my headlamp only showed a short distance ahead. The trail left by the victims' skis had been crossed over and over again by recent avalanches. Leaving the others behind to minimize risk and to watch, I skied over a dozen avalanche debris fields and farther into the valley. With my heart pounding as I nervously listened for approaching slides that could not be seen in the darkness, I carefully skied forward to see further into the valley. Some of the victims could be there, lying on the snow or partially buried.

When I saw the back of the valley in the moonlight and could see no obvious dark spots or possible bodies, I knew the time had come to leave the valley quickly. Terribly mixed feelings of abandoning the victims conflicting with the need to keep the rescuers alive ran through my thoughts as I returned to the waiting group of rescuers. Avalanches had run on all three sides of the valley. It was more than hair trigger, it was certain death. I was never more relieved than when I finally skied out of the steeper avalanche terrain and onto the rolling hills below.

The weather got worse. In the light of morning, we heard an Air Force Rescue Coordination Center helicopter flying in, following the highway as a reference point in the snowstorm. Covering our faces to protect them from the rotor-blast, we approached the helicopter after it landed in the parking lot and spoke to the pilot.

"We can wait a while for a hole in the clouds, but it doesn't look good," he said. "We can't fly anywhere in this without some visible reference."

Shortly, they would have to return to Elmendorf Air Force Base as the weather continued to worsen.

The victims' last hope was gone.

With somber hearts, we began the snowy drive home, but avalanches had closed the road between Anchorage and Girdwood. As we waited in Girdwood, we watched as naturally released avalanches ran down the mountain slopes and over the road. The whole region was drastically unstable. Later that day, after the Highway Department cleared the debris, we were able to drive the road back to Anchorage and home.

Days later, as the storm cleared, Mark Moderow, his avalanche recovery Border Collie, Hamish, and I helicoptered over the Taylor Creek Valley and saw that the whole valley had avalanched. Snow falling from the high peaks to the side of the pass had piled debris upon debris. The accumulated snow was so deep it took the melt of two full summers before all the victims were recovered.

Years later, as the avalanche school students skied into the valley, it was beautiful. The spring sun was warm. The snow was stable and a delight to ski. The avalanche class finished its studies and began the ski down the valley and back to the classroom.

Roni had skied in with our fifteen-month-old daughter Crystal in her backpack. Crystal would ride down with me as we skied back to the road.

Like I said at the beginning, backcountry skiing can be difficult. Trying to be careful with my precious cargo, I nonetheless buried my ski tips in an unforeseen snowdrift and fell, face first into the deep snow. My head was buried, but worse, the pack carrying Crystal had slid up from the force of the fall and was on top of my head! I was stuck. Crystal had been thrown partially from the pack, also burying her head in the snow, and flopped on top of my head like a freshly caught, twenty-five pound salmon.

I thrashed and rolled out of the powdery snow, and deeply concerned for Crystal, got her out of the pack and said, "Are you okay, Crystal?" As I cleaned snow from her sunglasses, got her hat back on, and brushed her off, I apologized. "I'm so sorry."

Except for a dazed and surprised look she seemed fine. At fifteen months, she was not yet talking and didn't seem to have much to say until Mom skied up behind us.

Suddenly, Crystal came out with a language all her own. Waving her arms and saying, "Bleah, bleah, blup, blup, bleah," she described to Roni in full detail what I had done to her. I'm not sure she ever forgave me.

PAPER, SIR?

Each year on the summer and winter solstices, the Mountaineering Club of Alaska sponsors a camp out on the summit of Flattop Mountain above Anchorage. Any number from a few to a hundred mountaineers visit and enjoy the views and camaraderie. The winter solstice weekend nearest December 21st is the least attended and often experiences temperatures well below zero as well as an icy climb to the summit.

One cold December, my friend Mark Moderow and I considered how to heighten the experience for those camped on the icy summit. "You know what everyone needs?" I hypothesized, "a tent door delivery of the Sunday paper."

Mark said, "The *Anchorage Daily News Sunday Edition* comes out around 3:00 a.m., which would give us plenty of time to climb the mountain in the dark and surreptitiously deliver a newspaper to each tent door or snow cave entrance."

Crampons squeaking on the dry snow and ice, we ascended the icy ridge in the darkness by the light of our headlamps. Trying to be as quiet as possible, we walked from camp to camp, barely able to contain our laughter as we thought of people emerging from their tent in the sub-zero morning light to find the current Sunday morning paper.

After the solstice camp out, MCA President Paul Denkewalter wrote in an issue of *Scree*, the club newsletter, that, "The paper delivery was a great idea, but where were the coffee and pastries?"

DEATH IN THE MOUNTAINS

"There have been joys too great to be described in words, and there have been griefs upon which I have not dared to dwell; and with these in mind I say, climb if you will, but remember that courage and strength are naught without prudence, and that momentary negligence may destroy the happiness of a lifetime. Do nothing in haste; look well to each step; and from the beginning think what may be the end."

--Edward Whymper, *The Ascent of the Matterhorn*

I have too many dead friends. The wild places of Alaska, while incredibly beautiful, can be very harsh and unforgiving. Whether it's a fishing boat sinking, a plane crash, an earthquake, or an avalanche, there are times of miraculous recovery and times of body recovery. Just living in Alaska is an adventure. But, in each incident there are lessons to be learned about priorities, values, and decision making. While you live, the learning never stops.

WHAT YOU DON'T KNOW CAN KILL YOU

The young man had no idea the snow he was playing on could kill him. Climbing up the steep, windblown gully, he stopped and began to slide down when the snow around him began to slide as well. Avalanche! The word inspires thoughts of tumbling blocks of ice and massive forces of wind driven, powdery snow. He died alone, buried under a few feet of snow compacted by the forces of the slide.

That afternoon, as I worked at Eberhard's Sport Shop in the city below, I heard of the avalanche over the local radio. Anchorage is only a few miles away from the steep and weather-pounded Chugach Mountains, it wouldn't take long to be at the site in hopes the victim would still be living. My boss relented and allowed me the time off.

I arrived at the parking lot of Robert Service High School, the rescue staging area, and met Doug Fesler, a Chugach State Park Ranger. Doug was a wild and wooly, highly intuitive force on the local rescue team, the Alaska Rescue Group. Asking permission of the local rescue commander for me to helicopter in with the rescuers to search for the buried man, though I was not yet an official part of the team, Doug said, "I've got a climber here who I think could be a help. I'd like him to come along." The commander said yes and we were flown in to the rescue site.

Darkness was growing as we began to organize and probe for the victim. It was difficult to see the full area involved. Doug viewed the avalanche path by searchlight from the helicopter to see how much remaining snow there was that might come down on the rescuers below. We continued the tedious work of an organized probe search. Teams of ten to twelve would line up in formation and at the command, "Down probe left," would push their metal probes as far into the debris as possible, hoping for that one "hit" that felt like a human body. "Probe up, down probe right," and the probes would descend just off our right toes. Over and over we would repeat these commands as we moved up the slide area.

Doug had come into the line next to me and kept mumbling, "I don't know why we're doing this, he's not buried here. He's up there in the debris in that gully."

"What are you telling me for? Go talk to the site commander," I said.

Hearing Doug's assessment of the situation, they quickly moved to probe the more likely area. They found the body shortly after. He had been dead for hours, likely suffocating shortly after he was buried.

I was young and had not encountered a dead body before. As I viewed it, I was deeply affected with the thought that what remained was only a shell, that the life that was the person was gone. We hauled the body to the helicopter on a sled and flew back to Anchorage. I was never the same after that first encounter with death in the mountains. I didn't know at the time that it was to be the first of many.

RIGHT PLACE, WRONG TIME

On the other side of the ridge in the Chugach Mountains connecting Flattop Mountain to Ptarmigan Peak, is a dynamic and beautiful ski mountaineering run from the summit of Peak Three. Starting right from the high ridge, it flows through bowl and gully and often has great snow conditions. It collects blowing snow from the slopes around it; great in the right conditions but potentially deadly if it overloads the slope with more snow-weight than the strength of the snowpack can handle.

He knew better. Bruce and I had almost died together in a similar situation on Mt. Silverthrone years before. My friend, Bruce Hickok and his companion on the climb, Geoff Radford, were not novices, or even intermediates who could be lured into a trap in their arrogance. They were highly experienced climbers and skiers, well equipped and trained, completely the opposite of the young man mentioned in the previous story. They even discussed the potential for an avalanche as they approached the slope that killed them, according to a survivor. Possibly their skills and experience lulled them into complacency. Maybe the storm was uncomfortable enough that they just kept their heads down and kept moving without consideration. Why they chose to ski that run on that stormy, drifting snow day is a mystery that they wouldn't live to explain.

Crossing the slope on the ascent, they were quickly enveloped in a large avalanche as the snow's strength failed. Buried in the debris, some of them had personal avalanche locator beacons, but they could not be found by the other members of their group as they searched the area.

In most backcountry avalanches, the best hope for survival lies with the team on the ground at the site. Not to be able to perform self-rescue is an almost incontrovertible death sentence.

TO THE SUMMIT

"To the summit, to the summit, to the summit," Ray Genet screamed, as Tom Ross and I listened in his equipment shed on Main Street in Talkeetna. Ray was explaining his philosophy of never turning back on a climb. Years before, he turned a client back who was a smoker and was struggling with the altitude. He told us that he made a decision after that climb to never turn back again. It would always be, "To the Summit," with him. Having turned a group back, Tom and I walked away from our jobs with Ray, shaking our heads in unbelief.

Ray was strong and powerful beyond description. Fun loving and charismatic, he enjoyed being known as "Pirate" and did all he could to keep the flamboyant image going. Legendary in the climbing history of Mt. McKinley, Ray was reported to have reached the summit of the mountain an incredible twenty-seven times and was one of the first to lead a guided climb of the mountain.

Visiting with him in Talkeetna after the climbing season of 1979, he told me he was headed for Mt. Everest in Nepal, the highest mountain in the world at 29,009 feet, later in the fall. He had a chronic cough, possibly the result of his many ascents of Denali and the thin, cold and dry air at high altitude.

Later, at Everest base camp, his philosophy of "To the Summit" would be tested when the team doctor told him he should not attempt the climb due to sickness. Ignoring the doctor's advice, he pressed on, reaching the summit on October 1, 1979.

Weakened by the altitude and sickness, Ray Genet and a German woman, Hannelore Schmatz, died on the descent after spending a night in a shallow snow cave above 28,000 feet. On October 2, 1979, the Pirate was dead.

Often, the highest performing climbers live by a code of "strength." To think about potential consequences is considered counter productive, even paralyzing. If they did not have this attitude of invulnerability, the risks would be considered unacceptable. It's why the seemingly impossible gets done. But, sooner or later, as with Ray Genet, human strength and determination isn't enough.

AFTER ALL THAT

Terry "Mugs" Stump was another superhuman climber, accomplishing the most radical and difficult climbs of the day, such as Moonflower Buttress on 14,570-foot Mt. Hunter, in the Alaska Range. Climbing the 5,000-foot high and extremely exposed rock and ice pillar, he set new standards for the word extreme.

The prolific, envelope-pushing climber, a man beyond strength, lost his life later, descending the South Buttress of Denali with two clients, as a corniced edge of a crevasse failed and buried him under massive chunks of ice. It seemed incredibly ironic that such a highly successful and experienced climber like Mugs would die in a simple accident.

Strength, and belief in your strength, are powerful tools and amazing motivators. Unbelievable accomplishments are made in the mountains and in life by people who are well-trained and unwilling to consider failure. But, human as we are, we are sometimes very frail.

IN THE REMOTE HIMALAYA

"There's been an accident! We need the medical kit. Now!" The voices below yelled up to us. We flew down the steep path, often just steps carved into the jungle forest trail, Sherpa Panzi carrying the large and unwieldy metal box that contained all our medical gear on his back and supporting all its weight on a tump line strap around his forehead. Being as careful as we could be, we ran down the slope and into the gorge below to find an emotionally-charged scene.

Steven (not his real name), a strong and capable middle-aged man, had fallen from the trail and was dead. One minute, he was enjoying traveling through the Himalayan foothills, and the next he was gone, killed instantly as he fell from the narrow path, over a cliff and into the trees below.

Expedition member Mark had watched it all happen. He was inconsolable.

Rappelling on ropes down the very steep, wooded slope, we sadly confirmed his death and prepared the body to be lifted with ropes back to the trail eighty feet above.

We were at least a week away from any communication with the outside world. Even if we were able to contact help, the Nepali Air Force would not risk other lives trying to fly a dead body out. The gorge was too narrow and help unavailable. We had no choice but to bury Steven along the trail.

The Sherpa people, our very capable guides, helped us dig his grave in the rocky soil along the trail. Intelligent and well educated, they nonetheless showed a superstitious fear of death as fetishes and good luck charms began to surface everywhere. Rabbit's feet and four-leaf clovers pressed in plastic emerged from pockets and hung on necklaces in an attempt to ward off evil.

A drizzly rain began to fall. After words and prayers were said and the group members settled in their tents, I hiked across the gorge to a waterfall I could see, hoping that some time alone cleaning up a bit might help me leave behind some of the trauma of the day. As I approached a narrow wooden bridge over the gorge, monkeys exploded from the brush, hooting and howling as they flung themselves down the trail and over the bridge before me. It was hard not to laugh at their antics as I bathed in the small, cold waterfall.

With the porters bivouacking in shallow caves at the bottom of the gorge, trying to stay out of the rain, we waited in a poor camp and through a sleepless night before deciding as a group whether to continue on with the trip or return to Katmandu. We all knew Steven would have wanted us to complete our goals and chose, as a team, to carry on with the expedition while one of the guides returned to report the accident to the Nepali authorities and Steven's family.

The terrain we traveled over the next two weeks through the remote Hongu valley, known for its Snow Leopard and Yeti sightings, was difficult. Large boulders covered by a deep and unseasonal snowfall made footing treacherous. As we plowed our way through the snowy boulder fields, we came upon an unexplained track. Some large animal, traveling on two feet had passed that way, leaving an animal-like, meandering trail in the snow. Recent snowfall had obscured the tracks, but whatever had made those tracks was tall enough to step out of the deep snow between steps. The trail wandered off toward the 20,000-foot Amphu Lapcha pass. A lone human would have been a strange sight in that desolate valley, and humans usually leave a more distinct, straight-line trail pointing toward their destination. We were left with more questions than answers.

The high pass we would traverse, Mingbo La, was at nearly 20,000 feet, and glacial, with a 500-foot ice wall descent to the more easily traveled glacier above the Khumbu Valley to the east. We lowered frost-bitten, tennis-shoe-clad porters by rope down the extremely steep and icy slope.

Flat ground never felt so heavenly. We realized, after walking for a time, that we had been intensely and anxiously focused on each step since Steven's accident two weeks before.

What causes these accidents? Ignorance can kill you, that's for sure, but many of these deaths were unnecessary. Steven's accident was just that, an accidental stumble at a bad moment, like tripping on a sidewalk and falling into traffic. Are there lapses in concentration due to weariness or the discomfort of stormy conditions? Or is it an attitude of invulnerability leading to a false sense of security? We don't know. All we do know is that a moment's inattention can put you in harm's way. Travel in mountainous terrain is like driving a car down a busy street; if you're not paying attention, dangerous surprises can happen quickly.

What's up with all this risk? Risk is unavoidable. The child who takes no risks will never ride a bicycle or camp out in the backyard. The teenager who will not risk will never drive a car or ask someone out on a date. The young person who takes no risks will never leave home to go to college or apply for a job. The adult who avoids risk will never try out a new occupation, start a new business, or commit to a marriage. Life is risky, from conception to our last breath.

Some take risks that are horrifying and unacceptable to others. Those risk-takers will tell you they have never felt more alive than when they were out there, on the edge. For an average person to attempt the exploits these highly trained athletes accomplish would lead to certain death.

Ultimately, it's about facing down fear. I don't know anyone who likes fear, but they do like the feeling of overcoming fear and not letting fear control them. The adrenaline rush of success in the face of fear is powerful.

Personally, I feel safer in the high mountain environment than I do driving a car down a busy street. But then, I survived many novice and intermediate mistakes on the way to that awareness and confidence. I don't believe in luck, but I do sometimes wonder why I survived when so many of my friends didn't.

I know this much. Death is a reality to be dealt with, not avoided. Each of these losses has marked me with a stronger sense of how frail we are as human beings and how mysterious the gift of life is in each of us.

"The race is not to the swift or the battle to the strong.... but time and chance happen to them all."

Ecclesiastes 9: 11 a.c. (NKJV)

PHOTOS

Jim on Denali, 1974

"Jim and Roni Hale on the summit
of Mt. Silverthrone, 1978."

"Jim and Roni at Home, 2008."

"Byron Glacier ice climbing,
1973." Photo credit Jim Johnson.

"10,000 feet above the clouds
on Mt. Spurr, 1973"
Photo credit Jim Johnson

"Silver Salmon and 'The World's
Greatest Super Cub'."
Photo credit Dr. George Hale

"Sunset on the Kahiltna Glacier"

"20,320-foot Denali, North America's tallest peak."

"Radio call from Denali's summit"

"Tent door view of 14,570-foot Mt.
Hunter. Kahiltna Glacier."

"Sgt. Bob Lapointe and Ken Wynne approach
Denali's summit in strong wind, 1976."

"Too fat to fly. Raven wing tracks in the
snow after attacking a food cache at
11,000-foot Kahiltna Pass."

"13,200-foot Windy Corner kitchen."   Roni Hale photo

"Hang-glider pilot Mason "Butch" Wade considers a narrow section of Denali's West Buttress." Roni Hale photo

"Pilot Bob Burns and Gene Maakestad haul a hang-glider through 18,200-foot Denali Pass."

"Looking from base camp to the summit
of Denali over 13,000 feet above." Photo:
Gary O. Grimm, Mountain Visions

"Pilot Ed Kvalvik assembles his hang-glider on Denali's
20,320-foot summit." Gary Bocarde photo

"Pilot Kent Hudson maneuvering above base camp on the Southeast Fork of the Kahiltna Glacier." Gary O. Grimm, Mountain Visions

"Climbing the crux pillar on the first continuous ascent of 600-foot Bridalveil Falls, Keystone Canyon, Alaska." Photo credit: John Dillman

"Roni Hale climbing the lower curtain of Ripple Falls, Eklutna, Alaska."

"Ice Avalanche crossing the one-half mile wide
Southeast fork of the Kahiltna Glacier."

"Glacier Pilot Doug Geeting gets low
on a fly-by. Little Switzerland."

"Roni Hale cross-country
skiing below the Moose's
Tooth, Ruth Glacier."

"16,400-foot camp on the ridge
of Denali's West Buttress."

"Landing on target at Mt. Alyeska." Roni Hale Photo

"Larsen Lake cabin." Roni Hale photo

"Jim and Mike Hale on Denali's summit, 2005."

"Climbing into the clouds on the North Ridge of Mt. Silverthrone, Alaska Range." Roni Hale photo

LANDINGS GOOD AND BAD

"Any landing you walk away from is a good landing."
                                        --Old aviation saying

"Fly an hour or walk a week," is one of the mottos of Doug Geeting Aviation in Talkeetna, Alaska. It's true. The combination of lakes, rivers, and swamps with the accompanying bugs, bears, and brush can keep anyone from their goals in Alaska. So, it follows that to get anywhere in bush Alaska requires a lot of time in the air in small, wheel, ski, or float equipped aircraft. Mostly it's a good thing. Occasionally, it's terrifying, even deadly.

The bush pilots of Alaska are legendary. Some have lived through a few crashes, others, like Cliff Hudson, the famous founder of Hudson Air Service in Talkeetna, never crashed a plane. In spite of the conditions he flew in, he was extraordinarily cautious. It takes time to learn the weather and the terrain. The old saying, "There are old pilots and bold pilots but no old, bold pilots," is true.

Glacier pilot Don Sheldon was a legend in his own time. He would often fly such long hours getting people on and off the mountain that he would hand over the controls to whoever was sitting next to him, whether they could fly or not, and say, "Just keep it pointed over there," and take a short nap. He would joke with passengers to the point of nearly giving them a heart attack.

Flying through One Shot Pass to Kahiltna base camp on Denali is always thrilling. The scale of the huge mountains is unimaginable and often takes people by surprise. The two snow-covered mountains on either side of the pass are steep and immense. To insure the ability to turn away from the pass in case of turbulence, pilots approach the pass from an angle rather than straight on. Don, knowing what he was going to do, flew straight toward the easternmost mountain to set up his approach to the pass. The mountains are so large that it seems you might hit them at any time, even though you are yet far from them. Don kept flying directly at the mountain wall.

On this particular trip, I watched from the backseat as his front-seat passenger, an older man on his way to climb Denali with his two sons was getting nervous. Just like a driver of an automobile who looks at his passenger and talks, rather than watching where he's going, Don ignored the approaching mountain, focusing entirely on his passenger and talking away like nothing was happening. "So, whadaya think about Alaska, great place isn't it?" Don would say as his passenger became increasingly nervous, his gaze flitting back and forth from the pilot's face to the fast approaching mountain wall. Back and forth his eyes would go as Don continued the incessant jabber. His whole body language grew increasingly tense, wordlessly shouting, *Don't you see we're going to crash into that mountain?* Finally, the passenger's jaw dropped in absolute terror as his hand lifted to point at the oncoming disaster. With a sly chuckle, Sheldon turned the plane slightly and slipped through the mountain pass at the best angle. As terrifying as it seemed, we were never close to hitting the wall.

As we unloaded at the Kahiltna Glacier base camp landing strip, located farther up the glacier and closer to Mt. Hunter than it is today, we heard a loud rumble from the 5,000-foot vertical drop avalanche chute above us. Not even taking the time to put his door back on, (doors are commonly and easily removed to facilitate removing packs and other bulky equipment,) Don yelled, "Get out of the way!" and gunned his turbo-charged silver Cessna 180 down the runway, beating the avalanche windblast by half a minute. The windblasts generated by these huge avalanches can travel a mile or more across the glacier and could have easily flipped his plane.

After letting the powdery, wind-blasted snow settle, Don returned to finish unpacking and retrieve his door.

Don was a hard act to follow. After Sheldon's death from cancer in 1975, Jim Sharp came on as a new owner of Don's old air service, Talkeetna Air Taxi, in 1977. He worked hard to live up to Don Sheldon's reputation. We began a life-changing interaction as he unexpectedly flew in to the Kahiltna glacier base camp before an approaching storm.

Bruce Hickok, Brian Okonek and I, all guides, had just returned from an attempted climb of the West Ridge of 14,570-foot Mt. Hunter. Stopped by difficulty just as the long, heavily corniced ridge blended into the mountain near the summit, we retreated in increasingly bad weather. From the base camp we heard the wind roaring over the summit plateau of Mt. Hunter, 7,000 feet above us. We were certain we would have to dig in and wait out the storm when, to our surprise, Jim Sharp landed in his Cessna 185 and said, "Hop in, let's go!"

We loaded the airplane with all our gear and the three of us, Brian in the very back, Bruce on his lap, and I in the front. This was in the days prior to the Federal Aviation Administration requiring that everyone had a seat to sit in. We took off in the overloaded plane, thankful to be headed home.

As we turned left from the Southeast Fork landing strip on to the main Kahiltna Glacier we were met by a fast approaching, 10,000-foot wall of cloud that went from the glacier surface to 15,000 feet! Trying to gain altitude, we flew circles above the glacier near Mt. Crosson. We were too heavy to climb above 11,000 feet. The storm kept coming, so we considered options. The first of our options was to land quickly and wait out the storm on the glacier. The second was to attempt to fly over Kahiltna Pass to the north and out to Kantishna, but Jim had no maps and didn't know the area. Taking the third option, Jim flew over next to Mt. Hunter to see if we could generate some lift in the winds flowing over Mt. Hunter. We were running low on fuel. As we circled above base camp, the clouds began obscuring the glacier below.

Suddenly, Jim took a right turn rather than continuing to circle to the left, and we found ourselves in drastic turbulence as we flew into the downside of a lenticular cloud, a lens-shaped cloud formed over a peak or a ridge in high wind conditions.

We were unable to see, the plane was being thrown about in violent turbulence and the pilot didn't know where we were. Trapped in the basin between 12,240-foot Mt. Huntington and the East Face of Mt. Hunter, the stall alarm shrilly buzzing, tossed like a leaf in the wind, we grimly held on. Massive walls of rock and ice flashed by in the patchy clouds. Seeing a particularly unique rock strata Brian, Bruce and I all exclaimed, as with one voice, "That's the West Face of Mount Huntington!"

"That pass, then that one," I directed the pilot to fly through.

Retracing our path, we returned to the skies above the Kahiltna base camp.

But, base camp was now totally "socked in" and we had no option of landing. Flying in total whiteout conditions now, we had no reference point until, just briefly, a very distinct, heavily corniced portion of the Northeast Ridge of Mt. Hunter appeared in the clouds just out the pilot's side window.

"That's the Northeast Ridge of Mt. Hunter!" I yelled.

"Are you sure?" Jim asked.

"Yes, I'm sure."

We knew where we were! Base camp was directly below us.

As the fuel gauges bounced on empty, Jim explained, "Here's what we're going to do. I can't climb above 11,000 feet. We're going to fly so many minutes and so many seconds directly toward Mt. Foraker, turn left and hope there's nothing out there we can hit."

Mt. Foraker is over 17,000 feet high. As we flew straight toward the mountain in high winds and thick cloud cover, the windows fully iced up from the inside. We were unable to see except where we scraped small holes in the ice with a credit card. The tension went beyond all bounds. Jim was flying with his right hand on the steering yoke and chewing on his left hand in overwhelming anxiety. We flew on in the cotton of the clouds. After a time of trying to imagine what it would be like should we survive the impact on the steep and massive mountain wall ahead, the adrenaline dissipated and we very nearly got bored.

We made the turn toward Talkeetna and down the Kahiltna Glacier. Jim asked, "Is there anything over 10,000 feet out there?"

Brian, who had climbed all over that area and knew it well, came unglued. "Yes!" he shouted, as he tried to claw his way out from the back of the plane in what looked like a desire to choke the pilot.

We began to slowly gain altitude in the thick, pea soup of the clouds, and finally, after forty-five minutes of extreme tension, broke out above the clouds at 15,000 feet. The Talkeetna ADF, or Automatic Direction Finder, pointed the way home. Jim flew as economically as possible to conserve what fuel we had left.

We watched closely as the ADF instrument switched from "to" to "from," indicating we were directly over the Talkeetna runway. Jim put the plane into a steep dive toward the airstrip over 14,000 feet below. We broke out below the cloud cover less than 300 feet above the ground, lined up directly on the runway ahead. We were going to make it!

As if we were destined to crash, the landing gear went into an uncontrollable shimmy as we hit the asphalt of the runway. Jim cursed his plane, pounding on the dashboard and commanding it to, "Give us a break!" We felt the plane would come apart. It didn't, thankfully, and we ultimately slowed to a stop before the Talkeetna Air Taxi cabin.

Later, Bruce and I found Brian contemplating our experience on the bank of the Susitna River at the end of the Village Strip, a short, dirt airstrip right in the middle of the Village of Talkeetna. Jim was sitting in his airplane hanger, vibrating from the stress. Bruce and I went to the Fairview Inn, trying to decompress the tension of the unimaginable experience before the drive home to Anchorage.

As I arrived home to my wife in the early hours of the morning, I woke her and said, "I died." The experience had a deep and profound effect on me. I found out that the storm had been so intense it blew screens from the windows in Anchorage. All that flying had been done in winds approaching 100 miles per hour.

NOT EXACTLY "SKY KING."

"One thing I cannot do," pilot Bob Burns said, "I cannot make a low altitude, up-glacier pass." He reminded himself of this fact and repeated it like a mantra as he prepared the underpowered Cessna 172 on wheels for the flight around Denali.

We were going to check out the mountain prior to the 1976 Hang Glider Expedition. Bob, one of the hang glider pilots planning on the expedition, was also a licensed private pilot and wanted to go take a look. At the same time we would check on friends who were climbing Mt. Huntington, a spectacular pyramid of rock and ice on the Ruth Glacier. We wanted to drop some ice cream to the climbers, and Bob knew it had to happen as he flew down glacier, as the little 172 did not have the power at higher altitudes to climb out going uphill.

Four of us—Bob, myself, my future wife Roni, and Ned Lewis—climbed into the small, single engine plane and took off from Merrill Field in Anchorage and headed toward the massive mountain two hours flight to the north. All went well as we slowly climbed to an altitude high enough to take a good look at the mountain. It was April and still very cold. The mountain was beautiful, and we enjoyed the spectacular views as we flew thousands of feet above the Kahiltna Glacier.

Slipping over the pass into the Ruth Glacier area, we looked for signs of our friends from Oregon on the glacier below Mt. Huntington. Roger Robinson, Cindy Jones, and their teammates were attempting the second ascent of the mountain by Terray Ridge, named for the famous climber who led the first ascent years before. We found them shuttling equipment up glacier to the base of the climb. Bob was very careful to only make passes over them as we flew down glacier. We dropped them their ice cream and a note wishing them good luck on the climb and began the trip home.

"One more time," Bob said with glee. He was having too much fun. He turned and passed over the team once again. Passing over their heads by a few feet, Bob then looked up from the snow below us and calmly said, "We're going in." In his excitement he had forgotten his rule and had made an up glacier pass!

"What do you mean we're going in?" I responded incredulously. He explained the situation as we looked for ways to turn or climb above the glacier. There were no options. To attempt a turn at the slow airspeed we were flying was to invite a stall and a possibly fatal spin into the glacier. The engine didn't have enough power at 7,500 feet to climb out. We were going to crash. Roni, who had taken her seatbelt off to take photos, scrambled to get her seatbelt back on. I held on to the back of the pilot's seat hoping not to hit the instrument panel when the plane hit the snow.

Finally able to fly no more, Bob flared the airplane with the same precision he would his hang glider and dropped us into the deep snow beside a large crevasse. The forward momentum rolled the plane forward onto its nose. The plane came to rest at an awkward angle, but we were upright and alive. With the plane perched on its nose, we looked down into the crevasse just outside the window and became horribly frightened as the fuel began to shift from the uphill wing tank to the downhill tank with a loud noise. For a terrifying moment, we thought the plane was slipping into the deep crevasse. Roni began to whimper. The rest of us, trying to think about options, in one voice said, "Shut up, Roni!" She took a deep breath and immediately calmed down.

The plane settled and did not slide. We carefully climbed out and surveyed the situation. After pulling the tail down to where the plane was more or less level, we could see the plane was unharmed except for a small dent in the engine cowling. Bob had performed the flaring maneuver so well the landing gear was not damaged at all. If we were not in a crevasse field and the plane were on skis instead of wheels, we could have flown it away! But, as it was, we were stuck.

We had landed so softly in the deep snow that the emergency locator beacon had not fired. After manually tripping the Electronic Locator Transmitter, or E.L.T., we realized we might need to spend the night in the plane and took stock. We had some cold weather gear, a few sandwiches and some water. We could survive for a while. We waited in the cold sunshine for who knew what.

After a few hours of waiting, the sun began to go behind a ridge, and we began to prepare for a long, cold night. Temperatures would likely go to -20 degrees or less. Just as we were climbing into the plane, we heard a yell. Looking down the glacier, we saw the Mt. Huntington team ski into view. We yelled and waved, hearing them yell in return. We found out later that they had not seen us at all and were thinking our yell was just an excellent echo.

Just then, we heard the "whup, whup, whup" sound of an approaching helicopter. They had been searching for us for hours, but the rock walls of the Ruth Gorge had played a trick on them, making the Emergency Locator

Beacon sound loudest at the snout of the glacier many miles away. The sight of the twin rotor Chinook helicopter rising from the glacier below was glorious.

The pilot hovered with the skis just touching the ground as we made our way through the deep snow and icy rotor blast to the helicopter. The crew met us at the end of a long tether, trying to keep us secure from any hidden crevasses. And, suddenly, we were gone.

The climbers never did figure out what had happened until we talked and compared notes later. This wheel plane just sat there, undamaged, in the snow. There was no blood indicating an accident. One week later they would watch from the ridge above as the plane was lifted out by helicopter, slowly spinning below the helicopter on its way to Talkeetna, where it would be inspected and flown home to Anchorage without need for repair.

The Federal Aviation Administration would later label it an incident rather than a crash. If emotional impact was any form of measurement, it sure felt like a crash to us.

GET RIGHT BACK ON THE HORSE THAT THREW YOU

In 1956, flying and shooting game on the same day was legal. Dad describes here a day in the life of a frontier hunter near Eureka Summit on the Glenn Highway.

"When my first son, Johnny, was six years old I took him moose and caribou hunting for the winter meat supply. He had a Daisy Air Rifle, which I had adjusted to his size by sawing off the back half of the stock.

"One fall morning we packed our hunting equipment into our station wagon and drove to a camp in the Nelchina area of Alaska. That afternoon we flew out in a float plane hunting with our guide, who was also the pilot.

"We flew by many lakes in an extensive area searching for game close to a lake on which we could land. At times, we landed and walked away from the plane looking for game. However, we did not find any moose or caribou to shoot. At dusk, we returned to camp for the night.

"The next morning Johnny was nauseated and couldn't keep anything down. We left him in camp shooting his BB gun at tin cans under the supervision of our guide's wife. We had apparently walked him too far the previous day.

"The guide and I, with a man named Tom Fink, who would later be Mayor of Anchorage, took off in a Cessna 180 on floats. We flew back into the hunting country at 400 feet and had just passed over a lake large enough to land on and over a small pond with a large beaver lodge at one end when the engine suddenly stopped. I was sitting in the right front seat with my left arm on the back of the pilot's seat.

"My first thought was, *Thank God that Johnny is sick today.*

"The plane was dropping rapidly, and the pilot was frantically flying the plane, turning toward the beaver pond, applying carburetor heat, switching fuel tanks, lowering flaps, pumping the primer and using the starter to try and restart the engine. We were dropping like a rock, and Mother Earth was coming up very fast.

"I thought we were going to hit hard. I cinched down my seatbelt as much as possible and thought that if I kept my left arm on the back of the pilot's seat I would get a fracture-dislocation of my left shoulder and of my lower back when we crashed. I put my hands on the instrument panel in front of me to brace myself. However, there was nothing but instruments covered with glass on the panel, and I thought I would ram both hands and forearms through the panel and damage them badly. So, I braced myself with my left arm on the back of the pilot's seat again. It's just how a surgeon thinks.

"The plane was diving toward the beaver pond and had turned 180 degrees. Just before we crashed into the muskeg beside the pond, the pilot, who, thank God, had kept controllable airspeed, pancaked the plane so the plane crashed on the bottom of the floats.

"We had enough forward speed that the plane skidded rapidly over the muskeg and then across the surface of the pond. As we started across the pond I thought, *maybe we are going to drown.* I saw the eight-foot-high beaver lodge coming, and we went right through it! We continued on another 150 feet through brush and into some trees and stopped. We were unhurt. We congratulated each other for being alive and then noticed that the fuel tanks in the wings had ruptured and gasoline was running down onto the hot engine. It was imperative that we get out before the tanks exploded in flames.

"I tried to get out of the right side of the plane, but the right wing had collapsed against the side of the plane, blocking the exit. I reached for my hunting knife on my belt to try to cut my way out when Tom Fink, in the backseat, turned across the seat and kicked the right rear door with both feet and the door popped off. We crawled out of the plane through the right rear door opening.

"We stood around and shook violently for a while and then noticed that the beaver lodge had exploded completely with no sign of it having ever been there. The plane's wings had collapsed down around the fuselage, the floats and their struts had been driven upward against the belly of the plane, the propeller had been windmilling in the sudden descent and had cut into the floats, the wing gasoline tanks were ruptured and the windshield had been knocked out of place.

"Eventually we stopped shaking, took our rifles and packboards out of the wreck and walked about 3.5 miles out to the Glenn Highway, where there was another lodge, and hitched a ride back to our camp.

"We boarded another airplane and took off from the lake, hunting again. Soon, we found two large bull caribou lying down on a grassy point extending into a lake from the eastern shore. There was another lake half a mile away westward, so we landed there and walked around the north end of the lake where we had seen the caribou. We could see the grassy point with two sets of caribou antlers showing above the grass, but their bodies were concealed behind the grass.

"Tom said, 'You take the first shot.'

"I estimated where the neck of one caribou would be and squeezed off a shot. The antlers dropped over and were still. Tom started to shoot, but found that the rear sight had been knocked off his rifle in the crash, so he said, 'You shoot mine, too.'

"I shot the second caribou in the neck, too.

"I have heard, and it has been my experience, that in a life and death emergency your mind is so concentrated that everything seems to happen in slow motion and is permanently stored in your memory. These events happened long ago, and I remember every detail as if it happened this morning.

"I have always been grateful that my friend, the pilot, didn't try to stretch a glide to try to reach the larger lake since it would have resulted in a stall-spin and probably a fatal accident."

I find it interesting that Dad's line of thinking just before the crash and mine during the Ruth Glacier incident were so similar.

LAST FLIGHT IN THE KITE

Soaring, for any length of time, is a hang gliders greatest joy. To break the bonds of inevitable gravity for a moment is wonderfully fulfilling. That's what we were after as we loaded up our kites on the cargo baskets of the helicopter and flew to the summits of the snow-covered peaks around Hatcher Pass in the Talkeetna Mountains.

The winds were just right, flowing in from the south-west, directly toward a broad face of the mountain, creating good lift.

Launching into the wind, we climbed above the mountain range in the steady lifting winds and greatly enjoyed the view of the valley below and all the way to Pioneer Peak in the Matanuska Valley many miles away.

Flying my new Seagull III hang glider and lying in a laid-out, prone position in my harness, I felt transcendent and above all the world below—for a few minutes.

An unexpected strong gust of wind hit me, flipping me over and turning me downwind, causing me to lose airspeed and control. I had no time to dive and try to regain the airspeed I needed to turn into the wind. I was at the mercy of the wind as I was propelled at frightening speed into the mountain slope before me. Thoughts of friends who had survived or died, depending on whether they hit soft brush or hard rock, flew through my mind.

The kite hit the snowy slope nose first, absorbing much of the energy of the crash into its aluminum frame. The wind then picked the kite and me up, rolling us up the hill in its power. Stopping upside-down and hanging from the control bar by a knee, I was unable to climb out of the kite or harness. One of my flying friends slid down the snow to me and helped me out of my predicament.

Disassembling the damaged kite, we carried it back to the ridge top to be flown out later. I hitched a ride on a passing snowmobile back to Hatcher Pass Lodge. Nursing a sore knee, I thought about the accident over and over, trying to find a way to have avoided the crash. I could not come up with any strategy other than not flying in strong winds in the mountains.

Good friends and better pilots than I had perished when unexpected turbulence threw them to the ground. I knew I was at a point of decision. Since this was all beyond my control and I felt I was pushing my luck, I made the choice not to continue to put myself in harm's way and sold the kite.

A BAD LANDING

Baranof Island, in southeast Alaska is rugged and beautiful. The original Russian capitol of Alaska, Sitka, rests on its western shore. Warm Springs Bay lies on the opposite, eastern shore. My brother John, his wife Deena, and friends Clark and Melinda Gruening, had partnered together to build a wilderness lodge there on Warm Springs Bay.

The bay was idyllic. Clear water, great fish, wildlife, and beautiful mountains created a scene beyond description. Baranof Warm Springs was a small gathering of older buildings capped by the larger store and natural hot springs bath house.

Caretaker Wally was an eccentric, but well loved host. When a boat would come into the bay, he would use a bullhorn to get the skipper's attention and direct them to a proper mooring. One day however, he was ignored. "To the boat just arrived, move forward to an inner mooring." No response. "To the boat just arrived, move to a forward buoy," he loudly directed through his open back window;

still no response. Taking up a high powered hunting rifle, Wally laid a shot across the boat's bow and commanded, "To the newly arrived boat, respond on channel six!" They promptly responded.

It was a unique and lovely setting. John and Deena loved to do long ocean kayak trips through the region, enchanted by the beauty and pace of life they found.

Transportation to and from the lodge was either by boat or float plane, usually from Sitka. Always having an uncomfortable feeling about flying with his wife at the same time, John would travel by boat when it was possible. But, in July of 1982, they needed to fly and chartered a Cessna 185 aircraft on floats. A powerful and efficient workhorse, the 185 is an often used and well-loved bush plane.

John and Deena, who was pregnant with their first child, waited out the weather for one day and took off from Sitka the following day in weather much better suited for flying. Crossing the rugged mountains at the center of Baranof Island, unexpected clear air turbulence seized the plane, flipped it on its back and slammed it into a mountain wall. Death was immediate for all on board.

Our family was devastated. Mom and Dad were beyond words in their shock and grief. Memorial services held in Juneau, Anchorage, and Oregon provided some comfort as friends and family shared the loss as well as the good memories and hope for eternity together. Most interesting were the many conversations John and Deena had with family and friends just before the accident. It was as if they knew this was coming.

Life went on. After a year had passed, Roni and I took our young family of Crystal and Jonathan to visit the lodge. Thinking I was pretty well through the grief of the loss, I was deeply surprised and troubled at the emergence of strong, sad feelings. While trying to help Clark with some painting projects, I became overwhelmed. "Clark, I'm sorry. I just can't do this." I said and then walked away toward the beach.

Sitting aimlessly, close to the waters of the bay, I was despondent with pain. Roni joined me at the water's edge and sat with me without speaking. I didn't know what to do. Surrounded by incredible beauty, all I could see was loss.

After a few minutes, a harbor seal slowly rose vertically out of the water like a submarine's periscope just a few feet away. He stared at me with soft and beady, brown eyes for a few moments and then slid, noiselessly, back into the water. We began to laugh. With that motion, the joy and wonder of the place came flooding in. Painful memories were replaced with the joy of life and the things John and Deena loved—their good friends Clark and Melinda and the natural wonders of Warm Springs Bay.

More than just accepting the tragedy we were now restored to the good aspects of their lives and reassured we will see them again. Filled with joyful memories of times together, life seemed to begin in a fresh and new way. Observing the grief and seeing that so few people were prepared to deal with it, moved me to pursue pastoral ministry training. What I do now as a pastor is directly related to that life and death moment. I was launched into the spiritual ministry through caring for our family and friends.

All these moments in life give you pause, reason to consider. Sometimes, in emergency situations, what seems like good luck and good preparation can bring you through to another day, hopefully a bit smarter and wiser. Sometimes, though, it's all over. Life as we know it on earth is finished. The question we all have to answer sooner or later is the one glacier pilot Don Sheldon would yell from his military jeep as he roared around Talkeetna, gathering up his clients, "Are You Ready?"

MORE THAN THE
MOUNTAINS

"Now I have given up everything else. I have found it to be the only way to really know Christ and to experience the mighty power that brought Him back to life again, and to find out what it means to suffer and die with Him. So, whatever it takes, I will be one who lives in the fresh newness of life of those who are alive from the dead."

Philippians 3:10-11, (TLB)

Exhausted, huddling in the small mountain tent on an icy platform at over 16,000 feet I felt, for the first time in my life, that I had come to the end of myself. As our team doctor and I worked at starting our stove to melt ice for drinking water, an overwhelming feeling came upon me that I would not live through this expedition. My heart felt as if it would burst, and my eyes filled with tears. I choked out the words, "I'll be right back," and stumbled out of the tent to hide my emotions from my friend.

The climb of Denali by its West Buttress should have been fairly routine. While statistics show that fifty percent of those attempting the climb are not successful in reaching the summit, I had ascended the route many times as a guide over the years with good success. The weather had been reasonable so far that climbing season and we had a well knit and capable group of climbers. As we flew to the 7,000-foot base camp on the Kahiltna Glacier, I was optimistic, but not in top condition as a result of a recent injury impeding my training. Climbing Denali however, is never routine. I should have known something was up when the trip began with a miracle.

Two days out of base camp my co-guide, Bruce Hickok, couldn't walk due to severe pain in his foot. We had no x-ray machine, but the two doctors in our group both felt he had broken a bone in his foot. I was concerned. A co-guide is an essential part of the team, allowing for greater flexibility and safety during the expedition. Sledding him back to the landing strip would take time and effort, and losing him as a support on the upper mountain could be serious.

I had heard of healing happening through prayer. I had no great faith, I was mostly desperate. I asked Bruce, "Would it be okay if I prayed for you?"

He responded with a frustrated and cynical, "Yeah, if it will make *you* feel better."

We went into our tent and I put my hands on his damaged foot, asking God to help Bruce. As I ended the prayer I looked over at Bruce, who had a shocked and amazed look on his face.

He had been healed.

Shouting with joy, he jumped out of the tent and began to wrestle and play with our startled teammates. This was certainly not your routine expedition experience. I had no idea how important it would be for Bruce to be with us later on the upper mountain.

As we continued the climb, I carried extra weight to compensate for one of the older members of our group, Hank Bahnson, a recognized heart surgeon who had attended medical school with my father. It felt as if Dad was on this expedition with me, and I dearly wanted him to succeed.

It seemed that wherever we went, energy-sapping rescues of others would occur. One such rescue began at Windy Corner, a narrow ledge of glacier ice wrapping around a ridge of rock at 13,200 feet. A European climber, traveling over the icy traverse without crampons stopped and visited with us a moment. Suddenly, he sat down on the steep ice and said, "My ankle is broken!" I was incredulous. It didn't seem possible. But, there we were, splinting his leg and ankle with foam sleeping pads and tent poles, wondering how to safely get him off the mountain. Helicopter rescue at the time was undependable due to the questions of availability and changing weather conditions. We were a long way from other assistance as Windy Corner was commonly reached from the 7,000-foot base camp on the Kahiltna Glacier, many miles away.

Running down the trail to an abandoned sled we had seen, I quickly returned to the team. Securing him into our climbing rope, we proceeded to sled him down the trail through the crevasse fields to the glacier below. Using men on each side to stabilize him, with others in front to pull and some behind to hold him back on the steeper sections, we descended rapidly to Kahiltna Pass.

Having been contacted by radio, Doug Geeting, a Talkeetna glacier pilot, flew his Cessna 185 ski plane to meet us, landing below 11,000-foot Kahiltna Pass. Waiting out the weather, Doug wrapped up in a sleeping bag in the back of the plane and read a paperback book, *Jungle Pilot*, while we watched for the cloud cover to clear enough for him to take off.

Only five hours after breaking his ankle at Windy Corner, the West German climber was being pinned and plastered at a hospital in Anchorage, over 100 miles away. Wearily, we trudged back to our camp 4,000 feet above and collapsed.

Endurance was not usually a problem for me. My life over the past years had been devoted to endurance sports. Competitive running, cross country skiing, and high altitude climbing had given me a tremendous base of conditioning. I had never experienced the total exhaustion of my resources and had always depended on them to be there and carry me through whatever might come. Now, though, I was about to see how truly weak I was.

Working hard to get the team to the upper reaches of the mountain, where storms, cold, and altitude can take a great toll, I grew increasingly weary in my body and spirit. Normally full of energy, I now struggled up the steep ice of the fixed lines, ropes attached, more or less, to anchors in the ice between 15,000 and 16,000 feet on the Headwall of the West Buttress.

I developed a sore throat, then a nagging cough, which robbed me of much needed sleep as it would constantly awaken me in the night. Despite efforts to heal and calm the problem, the thin, cold air at that altitude just caused it to worsen. I began to spit blood from a bronchial infection of some kind, shocking myself with the contrast between the bright, white snow and the fresh, red blood. I was deeply concerned, but for the clients' sake, not yet ready to quit the climb.

Our group continued on reasonably, but more surprises were coming. After camping at 17,200 feet, normally the highest camp, we began our ascent to 18,200-foot Denali Pass. The plan was to establish our high camp there, summit the next day and then descend the Muldrow Glacier route on the mountain's eastern exposure. Denali Pass is famous as a weather trap and a very poor place to be caught in a storm, as the wind amplifies its strength funneling between the peaks. Waiting on the plateau below with the last rope team, I watched the others travel up the very steep and icy, windblown traverse to the pass, ankles rolling and ligaments stretching under the weight of their 70-80-pound packs.

One of our team, Larry, had been feeling poorly that morning. He had not slept well and was feeling quite tired. Shockingly, sitting on the snow with his pack on his back, ready to go, he began coughing up bloody, brown foam. Pulmonary Edema is one of the most feared of high altitude illnesses. It's a leaking of bodily fluids into the lungs, and it can be deadly if not dealt with quickly. "Ooooweee," I called out to get the departing teams attention. "Hey! Come back," I yelled as they turned their focus toward us.

Retracing their footsteps, they were back to us within a few minutes.

"Larry's in trouble, it's Pulmonary Edema." I explained.

From there we began the evacuation of our sick friend and client.

Within our group were several capable climbers and guides. We decided that since the weather was excellent and the remaining group strong, I, my co-guide Bruce Hickok and our team doctor, Frank Hollingshead, would evacuate the sick man while accomplished climber and wilderness photographer Galen Rowell and the others continued on to establish our high camp at Denali Pass. It was then that we saw the flashing of light from the pass above. Something was happening—we could not know what—and someone was trying to get our attention! Another guide, Mike Graber and a client had climbed to the pass earlier in the day. We felt assured it was he who signaled us. Part of our team would be with him in a few hours. Thus, we began what was to become a two-pronged rescue.

Carrying our sick team member down the narrow ridge of the West Buttress was extremely difficult. The exposure can be extreme, with thousands of feet of icy rock walls below on either side. Lowering the patient on a rope was the only solution at times. As we descended into the thicker air at lower elevations Larry began to slowly recover his strength and ability to breathe. Below

14,000 feet, we were able to utilize a sled and move down the mountain fairly quickly. Finally, several energy-draining hours later, as we approached the 10,000-foot landing strip below Kahiltna Pass, Larry began to walk, slowly, on his own. He would fully recover later at sea level.

Brian Okonek, another guide and friend, and his two New Zealand friends had ascended to the 10,000-foot level from base camp, ten miles away, to assist us. Having heard the news on a Citizens Band radio we were using for communication, they walked through the night to meet us there and then return to base camp with Bruce and the sick climber.

While we were together, I told the group, "I have another problem. My K-boots have some moisture in them and are freezing on my feet at the higher altitudes. It takes a few people to help me get them off." Somehow, probably in a previous river crossing, water had gotten into the insulation of the rubber, vapor-barrier boots causing them to freeze into inflexible casts. It would take one person holding me by my shoulders and another pulling on the boot to get it off. In order to get them on again I would have to sleep with the bulky boots in my sleeping bag to keep them unfrozen.

Brian was the same shoe-size as I and was wearing an old pair of the same kind of boots. He offered, "Here, you take mine. As far as I know they're still good."

It wasn't the last time that a friend would come to my aid on this trip.

Soon, the doctor and I would need to return to the rest of the group at Denali Pass, over 8,000 feet up and many miles away. That, certainly, was more easily said than done. Normally, a group would take over a week to travel the terrain we had descended the previous day and night. Now, we were attempting to get back to our group in one day with little rest. We began to climb. One foot in front of the other, one deep breath per step; crampons and ice axes squeaking on the dry snow, the rope between us dragging on the snow like a multi-colored snake. Hours went by. Our sense of the incredibly beautiful terrain around us was dulled by exhaustion, dehydration, and the glare of the high altitude sun on the snow. At times, I would resort to simply counting the steps I would take, setting goals at 100 steps and then to rest. Each step required significant effort and thoughtful consideration of what it would take to get me up the next twelve inches. I used the Rest Step technique, where a climber locks his back leg knee at each step, allowing the muscles to rest as the bones and ligaments carry the weight for a moment.

We cat-napped at 14,200 feet as our mountain stove roared in the background, melting snow for water to refill our bottles and to cook soup. After the short rest, we continued up the steep snow and ice of the fixed lines, gaining new and dizzying perspectives as we ascended onto the upper mountain. Arriving at the ridge crest above the Headwall at 16,000 feet, we could again see Denali Pass. Aircraft were buzzing in the air and we watched, entranced, as a helicopter flew off from the pass and descended to the warmer, lower regions. Helicopter pilot and friend Ed Gunter had stuck an oxygen hose in his mouth to manage the thin air and climbed his Soloy helicopter to 18,200 feet to pull some injured German climbers off the mountain. We had no way of knowing our group above had been instrumental in the rescue of yet another climbing team, who, disoriented by fatigue and altitude, had taken a bad fall on the icy slope just above the pass where they had camped. That was the reason for the flashing distress signal we had seen the day before.

Total exhaustion came as we began to melt ice for more drinking water on the small, flat spot on the ridge. It had taken everything I had to get this far; I had no idea what I could do to continue. I truly had nothing left. Then, the feeling came. As I walked away from the small tent, blinking away the tears, I felt the spirit of the Lord impressing on me that I would not make it off the mountain alive. I sensed a choice, to descend and try to save my life, or to ascend and rejoin the group that was waiting for us above. The decision was clear to me. I needed to go up and care for the people depending on me. My place was with them.

Goodbyes are rarely easy and even more difficult when the one you love is not there to say goodbye to. In my heart, I said goodbye to my dear friend and wife, Roni, and prayed that it would go well with her. As I left that goodbye behind, something supernatural began. I felt a great peace. Energy somehow came to my body. I knew it was not my own. I had come to the end of myself, but God was just beginning.

The climb over the next mile of rocky ridge to our previous camp at 17,200 feet went smoothly. The weariness was gone; I felt no fatigue. I spent much of my time wondering what would happen to me, how my death would come. Would it be a crevasse fall, the sickness in my body? As we approached the camp, the weather began to change. We could hear powerful wind screaming through Denali Pass, 1,000 feet above. Would it be death in a storm? I felt no fear, only a childlike curiosity. The Lord was near, His peace was real, and all would be as it should.

Arriving at the camp, we talked with Terry "Mugs" Stump, a climber well known for his extreme risks in the past on difficult climbs. (Mugs later died on the South Buttress of Denali). As we visited, listening to the roar of the wind slicing through the pass above, he looked at me strangely and asked, "You're going up in this?"

The answer was simple to me, "My people are up there."

Traversing the difficult, iron-hard ice of the slope below the pass was complicated by winds blowing from our back so strongly that we were forced to walk leaning backwards at an incredible angle. The wind was so strong that, at times, when we turned sideways to the flow, we were unable to breathe due to the slipstream of wind around our bodies causing a negative pressure and sucking the air out of our lungs. Arriving at the pass in a ground blizzard of swirling snow, we headed for an old igloo that I had seen on a previous load carry to the pass a few days before. Thinking our group had camped there, we were disappointed to find the igloo abandoned and filled with rock-hard, drifted snow.

Our situation became more desperate. Unable to see more than a few feet in the drifting, wind-pulverized snow, we tried various ways of locating our group. We yelled and listened. Using the igloo as a reference point we did perimeter sweeps, a technique where one person travels to the end of the 150-foot climbing rope and stands while the other travels a circle around him at the other end of the rope. The noise of the wind was so loud that we had to yell from inches away into each other's ears just to communicate. Nothing worked. We could not find the rest of our team. We resigned ourselves to an attempt to dig some form of wind protection and camp.

One last time, we decided to yell together at the top of our lungs, and, as we listened, the faintest of sounds returned to us. Again we yelled and heard a louder sound through the blasting wind. Moving in the direction of the sound, we found them and their wind-flattened tents less than 100 yards away. They, being down wind, had heard our cry. Joining all their voices together in a resounding *Horton Hears the Who* chorus they made themselves heard. We were overjoyed to see them all, healthy and well, in spite of the wind that kept collapsing their tents, leaving them to lie in their sleeping bags under a shell of flapping nylon.

We waited for days for the storm to blow itself out. Unable to keep the stove running well during the storm, we became weaker from dehydration and lack of food. Time lost its meaning. Without physical activity, we felt weak. Completing our expedition goal of climbing to the summit of the mountain became less likely with each passing hour. Our objective now was to survive. Sleep came with difficulty, if at all. The freight train sound of the wind, coupled with the sharp cracking of the loose nylon around us, wore harshly on our nerves. Condensed frost showered us with each snap of the tent walls, causing our sleeping bags and clothing to become damp and clammy. Unless the storm broke soon, we would be in a serious predicament.

For three days the storm raged. I was in a state of eternal now, nothing else mattering but the immediate moment. All that mattered was making it through the next few minutes. Past or future were beyond thought.

"Listen, the wind is calming down a bit," someone said. Winds that we felt had been gusting over seventy miles per hour were now topping out around thirty-five.

"Let's get out of this tent and take a look around," I directed.

Shuffling around in the moderate winds exploring the wind blasted pass, scoured down to hard ice by the incessant storm, we looked to the south, where the previous storm had come out of the Gulf of Alaska, hoping to see clear skies.

Major high altitude storms in the Alaska Range are often preceded by certain cloud conditions. High, thin Cirrus clouds, together with an increasing wind and thickening layers of Stratus clouds moving in at lower altitudes, are common, telltale signs of a coming multi-day storm. What we saw, approaching rapidly from the south, was a 30,000-foot wall of dark cloud. It was time to leave.

Speaking to the group I explained our options. Reaching the summit was not one of them. Their one chance had been sacrificed as they spent the one good day of weather in the rescue of the fallen team a few days before. Our original design had been to traverse the mountain, to climb the West Buttress and descend via Karstens Ridge and the Muldrow Glacier on the eastern side. We could still do that or we could descend the western route we had just climbed to base camp and take a ski plane flight out to civilization. Whatever we decided, we needed to move quickly, before the next storm arrived. The decision was made to continue with the traverse and walk out into Denali National Park. We began to break camp.

Tired as we were from the effects of the storm and the high altitude, with winds still gusting around thirty miles per hour, we moved very slowly. We were very cautious, afraid of losing critical equipment in the wind. The temperature, with wind chill, was well below zero. Our fingers, exposed to the cold and wind as we packed our tents, began to freeze. I was shocked to find two of the heart surgeon's fingers frozen hard and white to the second joint. "Keep moving Hank, get your mittens back on and swing your arms back and forth, it will force warm blood out to your fingers." Hopefully, we would warm up as we began to travel down the Harper Glacier to the top of Karstens Ridge. Finally, we roped together and began the descent.

All during this time, I felt an incredible peace remaining on me while the certainty stayed with me that my future time on earth was extremely limited. God's strength still carried me. I could focus entirely on the needs of the group. A vague curiosity was still there in my thinking, *Will it be a fall on Karstens ridge, or a crevasse fall on the Muldrow glacier?* Then, at the very end of the descent, there was another possibility: fording the mile-wide braid of the icy and glacial McKinley River.

We descended to camp at 15,000 feet on the Harper Glacier within a few hours. We would need to rest here for the very difficult 4,000-foot drop down Karstens Ridge the next day. The extreme winds of the past three days had carved incredible clear ice sculptures in the surface of the glacier called Sastrugi. I had never seen the wind-scoured patterns formed in clear, hard ice before. It was a concern, as we needed perfect weather with little wind for a safe descent of the very exposed ridge waiting below.

As we pitched our tents on the hard ice, the wind began to pick up its intensity. Knowing the hazards of tomorrow and the fatigue of our group, I spent most of the hours of the night building and maintaining windbreaks around the tents, trying to keep them standing so that the others could rest. Amazed that I seemed never to tire in spite of little rest or sleep, I prayed that we would be able to escape the upper mountain the next day.

The wind continued through the night.

Never had I seen a more beautiful morning. "Thank you, God!" I exclaimed. Clear and reasonably mild, the day offered wonderful views of the Muldrow and Tralieka Glaciers winding their way dozens of miles to the foothills thousands of feet below. We broke camp and moved across the slope of Browne's Tower to the head of the ridge. There, the lead rope team encountered very strange and difficult snow conditions that stopped them in their tracks. The wind-deposited snow was too hard to kick steps in, but too soft to hold the teeth of our crampons without sliding. Chopping a staircase down the initial steep and frightening pitch with our ice axes was slow, but our only option. After experiencing a short and frightening fall, stopped by my ice axe digging into the hard snow, I descended the long, narrow ridge at a turtle's pace with the rest of the team.

The ridge was miles long and descended over 4,000 feet. At the end of a very long day, drinking in the thick air of lower altitudes, we rappelled down the final, snowy face of the ridge to the head of the Muldrow Glacier below. That night the storm we had seen coming from the pass hit the upper mountain full force. No one would move on Denali's upper slopes for the next two weeks. Had we not moved when we did, the chances of our survival would have been slim.

Though we had many miles yet to cover over hazardous glacial terrain and dangerous rivers, a thought was forming in my head that I might just live through this. Strapping on our snowshoes, we began exploring our way through the icefalls and crevasse fields of the Muldrow Glacier, seeking a route that would avoid the steep, avalanche-prone, glacial walls around us. Our goal for the day was to reach a safe area just above the Lower Icefall and then, the next day, McGonnagal Pass, and the ice-free ground for the first time in twenty four days. We covered the day's descent with little difficulty and, though I was still spitting blood, the hope of survival continued to grow within me.

Skirting around the water-filled crevasses of the Lower Icefall the next day, we reached the ablation zone, or the region of melting of the glacier, where we saw and heard running water for the first time in weeks. The smells of the rock, tundra, and flowers overwhelmed us.

The beautiful sights and sounds of running water sparkling in the sun amazed us. Packing away all our ropes, crampons, harnesses and snowshoes, we descended into a valley in the foothills where, that night, we slept on warm tundra and moss rather than the glacier ice and snow we had been accustomed to over the last weeks. A final twenty miles and two river crossings would put us at Wonder Lake and the park road, ending our efforts.

Emotional exhaustion was creeping over me. The thought that I need not worry about a crevasse fall or a fall from a steep ridge allowed me to begin to relax for the first time in weeks. The sickness was still with me, but the odds of recovery were now far greater in the warm, thick air of the lowlands. With the relaxation came emotional awareness and release. I felt this so very keenly when, one evening, camped on a small rise named Turtle Hill, and overlooking our final obstacle, the McKinley River, I heard a distant yodel. I immediately recognized it as belonging to a good friend and guide who occasionally worked with me, Vern Tejas. Having a friend for me there at that time caused the pent-up emotions inside me to break out. I babbled for hours about the events of the expedition. I'm sure Vern thought I might be losing my mind. He stayed on with us to help in the river crossing to come.

Early the next morning, when the river was at its lowest level due to the freezing at higher elevations, we began the crossing. The McKinley River is a mile-wide crossing of braided channels of gray, silt-laden, icy water. Most of the channels were only knee deep, but, usually, a few of the channels would be deeper, requiring teamwork to cross safely. Our most common method was to link arms with a group of two or three others, allowing the person upstream to take the burden of blocking the water pressure of the current while the others walked in his wake and supported him. The current was fast and would often wash the gravel out from under our feet while we were walking, causing us to walk and slide downstream at the same time.

Our packs, at this time, often weighed between seventy and ninety pounds, as all of our glacier travel equipment, ropes, harnesses, crampons, snowshoes, and sleds were added to the load on our backs. They actually anchored us, to a degree, by adding weight and stabilizing us against the river's force. Should we be forced to swim, though, they would become a true hazard. Our waist bands on the packs were off, the shoulder straps loose, in case we had to come out of them in a hurry. The water temperature was so cold, running from the snout of the Muldrow Glacier just a few miles above us, that numbness would overcome our feet immediately. People have been known to weep from the pain as ice-water-numbed feet throbbed back to life after the crossing. And, should we fall in, our whole bodies would become useless in seconds as the cold hit. The crossing is always tense.

Taking over two hours to scout and ford its many channels, the river was finally behind us, gratefully without incident. Using the last of our fuel for our stoves to start a large driftwood fire, we warmed our chilled bodies back to usefulness and dried our clothes. The last few miles to the trailhead at Wonder Lake were through beautiful spruce forest and bog. Moose browsed in the early morning sun that filtered through the trees. Insects hummed and birds sang. We had been in the high world of ice and snow, where even bacteria do not live, for what seemed like a lifetime. We were thrilled to enjoy these things as if seeing them for the first time.

The park road was our goal for the day where, normally, we would take a shuttle bus to the railroad station at Denali National Park Headquarters. Then we would board a train for Anchorage and the end of our trip. However, when we arrived at the road, we found it closed due to the storm and the late winter weather! It was the last straw for me. I was crushed. As Vern walked to the Ranger Station at Wonder Lake, I collapsed against the trailhead sign and slept. I had nothing at all left. I was done in. My team was safe. I no longer cared about anything. It was finished.

The rangers at Wonder Lake had mercy on us and piled us, and all our gear, into their pickup trucks to carry us out to where the buses could travel. Along the eighty-five miles of gravel road, near Polychrome Pass, my wife Roni, alarmed at what may have been happening to us and concerned about our delay in returning, met us. I fell into her arms, in awe that I had the opportunity to see her again and feel her arms around me. I later learned my co-guide Bruce Hickok and his girlfriend Sabina had tried to hike in to meet us, concerned for our well being in the storms. In their concern, they had brought a special treat for us—fresh broccoli! They also, in their hurried concern, had forgotten to bring any food for themselves! The weather was atrocious and the cold rain was constant. Their tent leaked, they got soaked, and got really tired of eating broccoli. After waiting for us in the bad weather a few days, they decided to give it up and go home.

As I threw my water-soaked and ruined vapor-barrier climbing boots into the trash can that night, I felt the despondency of exhaustion and sickness. The grace of God that had carried me through the last ten days was gone. I had such questions running through my mind. What had been my responsibility for my clients being in such desperate conditions? Was I responsible, in some way, for the frostbite that could have ended the heart surgeon's career? Would they have been in that situation at all had it not been for my encouragement and assurance of reasonable safety? I was assured that, except for the grace of God, our trip could have ended very differently. Thankfully, everyone was alive, without severe injury, and very glad to be on the way home.

The illness, however, did not leave me. I gave all my other guiding commitments to others and sat at home in our apartment, despondent. Unable to move quickly, exercise, or laugh without coughing so severely I felt my lungs would come out my throat I waited to recover, and didn't. Deep concern and many thoughts about my future were my constant companions. Something had drastically changed in my life. The reality of the presence and power of God during that time had totally changed my perspective on life itself. Some things were no longer important, others far more so. All that seemed to matter now was finding out more about this God who had so amazingly carried me through this experience. A part of me had, indeed, died there on the mountain and something of faith had replaced it. I knew life would never be the same for me, but I was unsure of what the future would hold. Would my health return? Would I continue mountain guiding? Could I resolve the questions I had about my responsibility for the clients in those conditions? What direction would I take? I did not know.

One month later, visiting in what we considered our home church in the village of Talkeetna, just below the mountain, I went forward after the service to ask Pastor Mike Stewart to pray with me concerning my need for healing. As friends gathered around me and laid their hands on me in prayer, I felt a deep, penetrating heat come through the Pastor's hand on my chest. I was wonderfully and miraculously healed.

Death to my self, or my self-sufficient pride, was terribly traumatic. I had never experienced total exhaustion before and the reality of coming to the end of myself hit me hard. Even more perplexing was the strange and supernatural power that carried me over the last week on the mountain and then healed me.

We all sometimes face overwhelming circumstances in life that bring us to the end of ourselves. It could be a health or financial crisis, the death of a loved one or the loss of a job. One thing had been made clear to me— that nothing can separate us from the love of Christ and that He allows us to come to the end of ourselves only to show us His grace and the power of His life within us. We don't need to be afraid. When life's circumstances become impossible, He is there to help us through.

The past year had been one of transition. Roni and I knew things were changing around and within us. In 1978 I had been gone, guiding in Alaska and Nepal, for over seven months. On one particularly long trip to the remote Himalaya, a client had died. Communication being so poor in those days Roni had heard that it was possible I was the one who had died. It was hard on her. It was hard on our marriage.

In years past, I had foolishly told Roni she would never replace the mountains in my life. At that time, I meant it. The mountains, with the peace and purpose I felt among them, were my God, my reason for living. Now, though, my perspective had changed. My view of life, the Lord, and myself had taken a great leap into the realm of faith and eternity. My values had changed. Now, all that mattered to me was finding out more about this God who had carried me and whom I had come to know in a new way. The desire to follow Him and shape my life to fit His values began to change everything.

ADVENTURE IS WHERE YOU FIND IT

"Everybody needs beauty as well as bread, places to play and pray in,
　　where Nature may heal and cheer and give strength to body and soul."

--John Muir

ARABIANS IN THE BUSH

Alaska is tough on horses. Although much horse-packing is done, especially for hunting, humorous stories abound regarding horses, bugs, brush, and bears.

My wife Roni had her dream horse, a powerful, red Arabian stallion named Boriffic. Bo, as we called him, was a show horse in a previous life and a bit of a sissy, though strong. When someone would ride his mare away from his corral, he would come unglued, trumpeting and blowing, making the most amazing noises. Often, people in the Village of Talkeetna would hear him doing his stallion scream and come over, thinking we had a grizzly bear trapped in our backyard.

Bo had a high play drive. We gave him an old inner-tube, which he would take in his teeth and begin to spin around and around, finally letting it go as he watched it fly fifty feet away. We made him a Bo Ball that hung from the branch of a tree. He would take the rope suspending the ball in his teeth and begin to spin it like an Eskimo yo-yo, stopping the spin with a front leg, reversing the spin, and then sending the ball flying with a well placed kick.

We kept him for breeding and riding, not for show, and decided we should try a cross country trip to our cabin near Larsen Lake, about fifteen trail miles distant. There were many well-used trails over most of the distance, but the last miles were rough. At one point, there was a power line that we thought would have some kind of access road beneath it. Wrong!

Downed trees and swamp were the norm under the lines. We picked our way through, being very careful to guard the horses' legs in the labyrinth of downed trees.

It felt like being in a descending elevator, as Bo's front legs suddenly dropped into swampy muskeg. Thrown forward onto his neck, I was completely off balance as he began to lunge, trying to get back to solid ground. It only took one powerful lunge to send me flying over his head in a perfect flying V! That's where your legs go straight up over your head in a V shape as you fly, upside down, over the front of the horse. Landing flat on my back in the swamp about ten feet in front of the struggling horse was a revelation. I was uninjured, no logs, no stumps, only soft muskeg. Afraid that Bo might panic in the midst of the patch of downed trees and hurt himself, I crawled quickly back to where he stood. Instead of running, he just looked at his mare, Princess, and did his stallion, "Hey Honey, look at me," nicker. I did have a gun, and for a moment was tempted to use it.

We picked our way through the brush and Devil's Club to the cabin for a well deserved rest. Two days later, we would complete a circle over some better trails and long stretches of road.

SEEN ANY BEARS?

The fishing boat *Ramblin Rose* dropped the four of us off at Oliver's Inlet, a bay just southwest of Juneau on heavily forested Admiralty Island. Our goal was to portage our boats and gear over a one mile neck of land to Seymour Canal, and explore the area over a few days using our ocean kayaks.

Finding an old, mostly decrepit but still useable, rail tramway built by the Civilian Conservation Corps in the 1930s, we loaded our boats onto the small railcars, pushing and pulling them toward Seymour Canal. The rail tram cars would occasionally jump the track and we would have to stop, resetting the cars onto the tracks. We often wondered if it would have been easier just to carry the kayaks.

Stopping for the night, as we unpacked our gear in a Forest Service cabin we had arranged to stay in, John asked, "Has anyone seen the cook kit?"

We searched and could not find the cooking pots. We had left our cook kits back in Juneau. My brother John and I had identical kits, and it caused a potentially serious mix-up when we both put our kits away, mistakenly thinking the other was packed in the trip gear. Scavenging, we found an old coffee can that would do for a pot, after we cleaned and sterilized it by boiling water in it for a while.

Heading out in the clearing, early morning mist, we were surrounded by large rafts of seabirds. John and a friend each had single kayaks. Roni and I were paddling a borrowed, double Klepper Foldboat. We were headed for Pack Creek and the bear observatory there. Ten miles or so of paddling and we were pulling our kayaks up on the gravelly beach.

Emerging from our boats for a walk on the beach, we carried a 12 gauge shotgun as bear protection. This area was well known for its large population of coastal brown bear that can easily reach eight hundred pounds or more. As we walked, we saw a very strange sight. An old man, wearing red suspenders and a weather beaten and ancient straw hat with the crown painted International Orange above the brim, approached. He carried no gun but only a walking stick.

"See any bears?" the old man cackled. "Lots of bears around, see any bears?" It turned out the old man was Stan Price, who had sailed to the Pack Creek area in the 1920s. He had been around far longer than any of the large brown bears and had known and named them all as cubs. If a bear got feisty he would just give it a sharp rap on the nose with his stick and that would settle it. He invited us to his weathered and moss covered houseboat to meet his wife Esther and to see his deteriorating old sailboat and garden.

The garden, filled with luscious and juicy, red and yellow salmonberries as well as other vegetables, was protected by a strand of electrified wire. Stan told us of the many times he would see a deer jump the fence to be protected from a bear, and that the bears respected the electric fence. Later, we would take this information and use it to protect cabins in the interior of Alaska from marauding grizzly and black bear.

George McCullough, a long time miner in the Cache Creek District of the Alaska Range, had had great difficulties with bears, including one instance in which a grizzly pulled an entire wall off a cabin. He used the electric fence to protect his buildings and had no more problems.

The CCC, or Civilian Conservation Corps had built a good sized, metal tree house directly over Pack Creek in the '30s. We nervously walked the brush-hidden trail to the Observatory and climbed quickly up the ladder as fishing bears watched us from the creek just a few feet away. After a few hours of watching the twenty-plus bears nearby, we hiked back to our kayaks to launch out for our night's camp.

The richness of life in the ocean is awe-inspiring. Each moment was filled with new birds, eagles, sea lions, and whales. During our last day in the area, the wind picked up strongly from our backs as we paddled back to the head of Seymour Canal. Holding on to each other's boats, we made a loose tri-maran of the three boats. Stretching a tent-fly between two kayak paddles, we made a large, downwind sail. The wind caught the sail and propelled us up to ten knots an hour as we guided the boats with our rudders. Traveling so fast without paddling was amazing and fun.

Silently skirting a small island, we surprised a large group of sea lions with our sudden appearance. They avalanched into the water around us, barking and growling excitedly at the strange, square shaped beast that had appeared.

TALKS WITH CARIBOU

The fall season north of the Alaska Range is a magnificent, multi-colored, multi-sensory experience. The smells of the ripening berries and damp tundra, the colors going rust, red, and yellow as the frosts turn the leaves, contrasted by the bright white termination dust snow blanketing the 20,320-foot mass of Denali are nothing short of spectacular. The termination dust, so called by Alaskans, signals the end of the short Alaskan summer as it creeps lower on the mountain slopes day by day.

Members of my family won a weekend at Camp Denali near Wonder Lake but were unable to go. My wife and I, and our two young children, Crystal and Jonathan, needed some time away, so we jumped at the chance.

After a long bus ride from the Denali National Park entrance and a good night's sleep, I awoke and took a walk around the lodge. Soon I was running over the tundra, reveling in the beauty and freedom, amazed at every turn with new vistas. Noticing a large bull caribou rack sticking up through a tangle of brush, I slowed to a crawl and began to see how close the animal would allow me to approach. The caribou are protected in the park and fairly comfortable with people.

Talking gently and moving slowly, I came to within fifteen feet or so and sat down as the large bull continued to lie in a small clearing in the brush. Time passed as we sat together, enjoying the moment. Finally, I left the bull resting and continued on with my walk.

WILDERNESS CABIN

Of all things Alaskan, building a remote cabin is what many people think of as an ultimate wilderness experience. I had no idea how fulfilling it would be. We had purchased land from the State of Alaska for a cabin site near Larsen Lake, a few miles east of Talkeetna at the base of the Talkeetna Mountains. Choosing the land from a map and a tree-top over-flight with Don Lee in his Super Cub started the process that would take years to complete.

Finding the way into the area was a challenge in itself. There were no roads or trails. Cross country ski trips with sled dogs and trial and error attempts at finding the best trail opened the way. The process of building began as we cleared trails and freighted materials in by snow machine.

We would build a small log cabin for our family and friends to use as a getaway. None of us had a lot of money so it would have to be done as inexpensively as possible. I girdled the chosen spruce trees, cutting back the bark at the base of the trees to cause them to die and dry where they stood, making them much lighter to log and build with.

Most of the logging I would do alone as I had few others available to help on my schedule. The logs dried over the next year and were very reasonable to handle. Using an ancient $350 Ski-Doo snowmobile as both freighter and logger, I skidded the dried logs to the building site on a make-shift sled. Good friends Dan Buchanan and Don Lee pitched in to get the structure up, and we were off to a good start. The whole family pitched in with the work. Roni would scrape and paint the old windows we so carefully freighted in and the kids would peel bark from the spruce logs and pound spikes to hold logs in place.

A project like this is never finished. There is always something more to do—an improvement to make here or there. Creating a home in the woods, far from the nearest neighbor, was one of the best times of our lives. The friends, the family, and the fun we had on holidays all made those years very special. Ridiculously fun games of hide and seek and other diversions kept us laughing.

After building rope swings and a tree house on a hill behind the cabin we named Eagle's Nest, we left a foam mattress in the tree house. A visiting bear climbed up and into the tree house when we were gone and shredded it, leaving only small scraps behind.

Often I would go out alone, just for the quiet. It was a place of restoration, healing and vision as I enjoyed the deep peace. We heard and saw otter, beaver, moose, bear, owls, and wolves as we traveled the miles in and out. A solar-powered electric fence was wrapped around the cabin when we left. Though we saw signs of bear coming up to the windows, we never had a break in.

Walking the trail to the creek, where we would get water, we would often see bear territorial markings high on the tree trunks. Bears wanting to make their size and presence known would reach as high as they could on the tree trunks, scratching off the bark in a show of bruin one-upmanship. Though the bears were always around, we rarely had an encounter.

Roni and I would often spend a few days away from work and the kids just to keep the Jim-and-Roni part of our lives refreshed. In the midst of raising small children and the work of the church ministry, we found the quiet time away essential to maintaining our sanity.

MIRACLE ON THE COLORADO RIVER

Lessons learned through experience often find their purpose fulfilled at a later date. This lesson, learned in 1976, had life-saving impact on a young man's life over twenty years later:

I was tired and not paying attention. As our rope team traveled in a stiff breeze around 13,200-foot Windy Corner, on the West Buttress route of Denali, I thought it would just be routine. We were traveling an established trail after all. One foot in front of the other, head down to protect against the wind, we plodded on. In a heartbeat, I was into a crevasse up to my waist, hanging from my external, aluminum-frame backpack, my legs swinging freely in the darkness below. In my weariness, I had not properly prepared my Jumar rope ascenders. If I took a free fall into the crevasse, I would be unable to climb out. Looking down into the inky blackness of the deep crevasse, I was terrified.

To complicate matters, my rope mates were unable to see or hear my yells to go into ice axe self-arrest and anchor the rope. I was just over a small rise, and the wind was drowning out all communication. Each step they moved closer to me guaranteed a longer fall into the crevasse, possibly pulling them in behind me as I fell.

Looking over my shoulder, I saw the second man on the rope, Dutch climber Marcel Terbeck, begin to crest the rise. I yelled into the wind, "Marcel! Crevasse! Self-arrest!" My screaming at him to get the rope tight and go to self-arrest had no impact on him. In the wind and weariness he couldn't fathom what was happening.

Ken Wynne, an Anchorage dentist, was on the other end of our three-man rope. Coming over the rise, he quickly saw what was happening, threw Marcel back down the hill to tighten the rope and strode forward to grab me by the top of my pack and pop me out of the hole. As far as I was concerned, that was the best extraction Ken ever performed.

A disaster had been averted. I was thoroughly humbled by the fact that I had walked into that trap unprepared. My rescue gear would do me no good at all if it was packed away and unavailable.

Fast forward to the Colorado River, twenty-two years later—I was preparing for an easy afternoon float with my kids and some of their cousins and friends. I always took rescue gear in my kayak, but rarely needed it. I looked at my rescue knife, the rope throw-bag and carabiner snap-link. Would I need them? *Not likely*, I thought, as the Colorado River passing through Grand Junction, Colorado was an easy float with little to run into. I almost left them behind, but remembered my experience on Denali all those years ago and put them in the boat. *If you carry it, you won't need it. If you don't carry it, you will.* I reminded myself.

The afternoon was idyllic, the weather was beautiful and the sun was warm. As we floated peacefully under high cliffs, Great Blue Herons fished the shallows and then rose in majestic flight on their long wings. We met a group of three young people on the river, floating in inner tubes tied together, with make-shift oars made out of lumber. Usually this wouldn't present a problem, and we watched them float by, waving and saying "Hi." They didn't know about the logjam piled up on an old concrete bridge pier downriver—easy to misread, even if you're looking.

They weren't looking. From two hundred yards upriver, too far to prevent it from happening, we watched them be swept into the logjam by the strong current. The river was still running high in mid-June, and cold in spite of the warm afternoon. We watched, helpless as they approached the brush and trees pinned to the upstream side of the old bridge pier.

Warning my fourteen-year-old son, Jonathan, and my nephew, Cody, I told them to watch carefully to see if everyone made it out as they hit the piled up trees and brush. One...two, we saw them pop out. One was missing. I prepared the boys, telling them that we were about to be involved in a rescue as we paddled to the river bank near the log jam. As we beached the boats, we could see one young man trapped in the logjam with his head just barely above the water as he screamed for help.

The water broke over his head in waves as I gathered the rescue gear I had almost left behind. He was terrified; we were all overwhelmed, he in the water, and us in the deadly potential of the situation. This was as serious as it gets.

Against the sound of the breaking water, we took turns yelling encouragement, "Hold on, we're coming!" The noise was so great he was unable to hear us. Praying in his terror, he said, "Help me God, I'm too young to die this way." With his multiple piercings and green Mohawk he wondered, *Who would want to save this kid?*

Strapping the rescue gear onto my life-vest and climbing onto the sit-on-top kayak that Jonathan had been using, I paddled behind the bridge pier, almost capsizing in the turbulent current. Allowing myself to be swept up into the logjam from below by a strong eddy, a swirling current found downstream of an obstacle, I pulled into the mass of sticks, twigs, and logs. Climbing carefully onto the log jam, I threaded my way through and over the top of the piled-up debris, often called a strainer. Bending down, I reached for the young man and tried to lift him from the current. He was held too tightly by the rushing water. I couldn't move him.

Talking to him in hopes of calming him down, I said, "Hi, I'm Jim. What's your name?"

"Scott," he replied, between screams of, "Help me, I'm going to die!" One of the makeshift oars had gone down the sleeve of his T-shirt and twisted in the current, making a tourniquet and cutting off the blood supply to his arm. Taking my rescue knife, I slowly and carefully cut the shirt away. The shirt had been twisted so tightly it was very difficult getting to it with the knife without cutting Scott. Finally, the oar was free.

I could now thread the rope around his chest and try to rig a pulley system to lift him from the river. I tightened the chest rope, tying it off, and then looked for the best anchors I could find to build the lift. All the branches above Scott were too small and too weak to make good anchors, but I did what I could as the water roared through the trapped brush. I tried to lift him out using the pulley system. His head came slightly out of the water, but he was stuck tightly

This was not good. I prayed as I tried other ways to help, telling God I would rather not be there if I was just going to hold Scott's hand while he died. He kept screaming. I finally had enough. As he screamed, in great fear, "I'm going to die," I told him, "Yeah, you might, but shut up!"

"Oh," he said, "Sorry," and he quieted down.

I was out of options. Scott was tiring quickly and there was no other help available. I thought we would need a power boat and a chainsaw to make any headway, but power boats were hardly ever seen on the river and chainsaws not too likely. Certain I would witness Scott's death without a miracle, I desperately prayed, "Lord, if you're going to show up, now would be a good time!"

Time seemed to slow to a crawl as I stood helplessly waiting. After a few long minutes however, I was shocked to hear the buzzing hum of a small engine as a Wave Runner personal watercraft came alongside the logjam. The other tubers in Scott's group had washed down to the Corn Lake State Park boat launch and had yelled for help. Seemingly coincidently, a man was playing with his watercraft just off shore. In all our years on the river, we had never seen a powerboat or watercraft on this section of the river. The other members of my group told him the situation and sent him up the river to find us. He hesitated at first, understanding he was not allowed to go upstream from the boat launch. But, he had reconsidered, and now, here he was.

The current and waves were too strong for him to pull too close. After trying to yell over the river and engine noise and then communicating with him in gestures I unwrapped my rescue rope from the logs and threw the driver the rope and carabiner snap-link to attach to his watercraft. Trying to keep his machine steady as it bounced in the waves, he steered with one hand and caught the rope in the other, clipping the rope into a handle on the side of the watercraft. I motioned him to go straight upriver as fast as he could, and he took off. Explaining the last ditch plan to Scott, I yelled, "Take a deep breath!" as the rope came taut.

Scott went under as the Wave Runner tried to pull him from the logjam. As the small machine strained against the current, I could just barely see the light reflecting off his body two to three feet below in the murky water. It wasn't going to work. The small machine didn't have the power to overcome the swift water. I waved the driver off to the side to try to pull across the current just as he was washed to the side by the turbulent waves. He hit the throttle, and the machine and rope strained against Scott's trapped body.

Suddenly, like a white torpedo, Scott shot through the water and was free. Better yet, he was alive! I watched, breathless, as he slowly climbed and was hauled by the driver onto the Wave Runner, completely naked except for one sandal. The branches of the trees in the log jam had held him trapped by his clothes, which then ripped off as he was pulled out, releasing him from certain death. "Yes! Thank you God!" I shouted in great relief.

As the driver headed off downriver to the boat launch and a waiting ambulance, I gathered my wits, tried to breathe and looked at my situation. *How am I going to get off this log jam?* I thought to myself. The current was deadly from above, the circling eddy current too strong to escape from below. My kayak had floated free and was no longer there. I would have to jump as far as I could into the current on the side and hope it would carry me past the debris.

I jumped. It worked and I found myself floating free down the Colorado, resting in my life-jacket. Jonathan paddled alongside to tow me to the beach at Corn Lake, where I shook hands with Scott and told him, as he retched river water, "I'm so glad you're alive!" The paramedics bundled him into the ambulance and off to the hospital. As the ambulance left, the adrenaline wore off. I walked a short distance upstream and emotionally collapsed in great sobs of relief.

A few years later, I would have the honor of officiating at Scott's wedding. There was never a more happy and filled-with-life celebration. And all because of a knife, a rope, and a carabiner that almost got left behind.

TRAPPED IN THE OUTHOUSE

It was a great outhouse. A large window, with curtains, looked out into the Birch forest behind our Talkeetna house, and a caribou antler in velvet graced the door as a handle. Light came in through the corrugated fiberglass roof, and a Styrofoam seat made it warm on those cold winter mornings. The interior was painted a clean white and the floor was covered with an old piece of scrap carpet. It even had a guest book!

It was a symbol of marriage promises fulfilled. Roni had come to the end of her tolerance with the old outhouse. We had very little money, but I had promised her this would be the last summer she would have to put up with the old outhouse.

The old one was *old*! Made of scrap, unpeeled spruce poles, rough lumber and asphalt roofing, it was slowly falling into its hole. Moss grew under the seat boards. It was dark inside. As you sat in it, you had the distinct feeling you were going to fall over backwards.

But, as I looked at our money—or lack of it—at the end of the summer, I told Roni I didn't think we could afford a new one. A marriage crisis ensued. "You promised!" she wailed. Well, I had promised. I borrowed some money from my understanding parents and began construction. This was going to be, not just an outhouse, but a great outhouse.

Visitors enjoyed it. Books and magazines began to show up in it. A good friend, Cindy Jones, provided a guest book. People signed in from far and near.

One cold and snowy winter day, while the house was filled with visiting friends, one Peter Chapman decided to visit the facilities. He enjoyed the décor and flipped though the guest book, recognizing the names of many old and dear friends. Upon finishing, he pushed on the door only to find it locked tightly.

He searched and searched for a way out. He yelled for help, but his voice could not be heard over the voices of the noisy crowd. He looked at the window, thinking he might have to break out. But, as a New Englander, he recognized those small panes in the window were old glass and precious. He couldn't do it. He yelled some more.

Half an hour later he was still locked in. He resigned himself to wait it out until someone would recognize his absence and come to his rescue. Or, he hoped, someone else might need to "go" soon and open the door to his escape.

He sat and read through the guest book once again, and, after a few minutes noticed an asterisk and note on the inside, front cover, that said, "Oh, by the way, if you find yourself trapped in here pull the white string!" He hadn't noticed the door was latched from the outside as he entered. Upon his return to the house, it was added insult to injury when he saw that no one even noticed he had been missing.

A TRAIL OF BANANAS

Loading up the two-seater Super Cub with all our hunting gear and a five-gallon bucket of bananas to give Larry and Naomi Rivers as a gift, we headed for the Brooks Range hundreds of miles to the north, to hunt caribou.

Flying with bush pilot legend and in-law at the time, Don Lee, was always exciting. Landing on a rocky beach with one wheel in the water or flying at treetop level to avoid weather were nothing to him. We were out to get a caribou or two.

Approaching the hills south of Fairbanks, the cloud cover dropped to the ground. Flying in and out of the clouds, Don finally put the small plane down on the highway and taxied the plane over the summit road and down the other side, stopping in a line of traffic just like a car, as a highway construction flagman held up his STOP sign. We weren't going to let a little weather stop us.

North of Fairbanks the cloudy weather continued. We took the other approach, climbing above the clouds for a time. That was beautiful, but with no flight instrument information available at that time, we would soon need to see the ground again to figure out where we were.

Spotting a good-sized hole in the clouds, Don spiraled down toward the ground. There, shining silver in the distance was the Trans-Alaska Oil Pipeline, transporting crude oil to the terminal in Valdez. A large river was nearby, and diving the plane at the bridge over the river, we read the sign that said "Yukon River." We knew where we were.

Turning upriver, to the east, we headed for Ft. Yukon, not far from Larry and Naomi River's Dall Sheep hunting camp. All along the way, we snacked on the bananas we planned to give to the Rivers and their hunters. Throwing the peels out the window over the vast and trackless Alaskan Interior, we felt like Hansel and Gretel leaving a trail of bread crumbs to follow home. Small chance of that happening! Dealing with strong tailwinds that tried to blow us off course, we managed to find the right valley and drop in on our friends. Larry and Naomi had a comfortable tent camp set up, and that night we feasted on delicious Dall sheep back-strap.

Flying into the caribou area, Larry dropped his son, Shane Rivers, off to hunt with us. Following Larry's Super Cub onto the short strip, Don landed short and in the creek just before the gravel bar. After we bounced out of the creek and stopped the plane Larry walked over, looking at Don and saying, "Don…in the creek?" We set up a small tent and rested until early morning. Don, a master at a fast get away, in the morning swallowed a package of dry, instant oatmeal and a spoonful of instant coffee, following them up with some hot water. Breakfast took about thirty seconds.

Stalking a small herd of caribou, we finally shot one, butchered him out and headed back. Weather again forced us down onto the Parks Highway on the return trip. We pulled the plane over onto the side of the road and pitched our tent under the wing in heavy rain until the weather cleared enough to reach Talkeetna.

I'm sure more than a few people wondered what we were doing, but its all part of Alaskan flying, where there are very few landing strips, and you make do as best you can. By the way, if you find any banana peels south of the Yukon, you'll know where they came from!

A BIG MOOSESTEAK

The piles of freshly butchered moose meat grew on the kitchen table. We could hardly wait to fry some up and have a meal. Michael, our four-year-old son, watched with growing concern as we put meat in the skillet and prepared to eat.

A few hours before, we had happily received a phone call that told us we had a moose waiting for us at the railroad crossing below Ski Hill near Talkeetna. It had been hit by a train. The Alaska Railroad had partially butchered the animal and had brought the carcass to town with others, who had also been killed on the railroad tracks. Many such moose are killed in high snow years, as they find the railroad tracks easier to walk on than the deep snow and they have difficulty avoiding the trains. The meat from the animals is not wasted, but is distributed to people who sign up on the "Road Kill List" that includes both highway and railroad kills.

Michael had watched, in the glow of the headlights of our truck, as I carried knife and axe through the light snowfall to the carcass and began to cut up the young moose. "But Moosies are my friends!" he wailed. Roni explained to him the facts and feelings of the moment. Back in the house, young Michael found himself unable to eat when piles of freshly cut meat filled the table. After we cleared away the piles of raw meat Michael enjoyed his moose dinner.

Our kids, Michael, Jonathan, and Crystal had lived closely with moose their whole, young lives. Moose, at times, lived in our yard in tough years. One moved into our woodshed in an especially snowy season. When needing to leave the house, we would open the front door, look carefully in both directions and head for the car, hoping to avoid a head-on meeting in the deep trench of our snow-shoveled walk.

In one particularly deep snow year, the kids had fed the moose tree branches from the top of seven-foot high snow banks. One moose we called "Patch" due to a healed injury on his hip, probably from a close encounter with a train. He lived in our driveway for weeks.

It had snowed over twenty-two feet that year, with no melt. Roads were blind mazes with seven-to-eight-foot walls. The moose were starving and unable to get to their normal food of brush and branches due to the deep snow. Sometimes they couldn't move at all. Compassionate residents used their chain-saws to fall trees near the weak and trapped moose. Others tried to feed them hay, not realizing that a moose's digestive system goes through a complete change from the summer swamp greens to the winter's twigs and branches. They could starve to death eating hay. This was a dangerous situation. Starving and harassed moose can be aggressive to the point of being deadly.

People were being attacked. They had to carry high-powered hunting rifles when going to check the mailbox at the end of the driveway. The people of the community had some very close calls that year but, thankfully, no humans died.

One friend, Anita Ford, was attacked and injured on her way to church. She was saved from being killed by Delores McCullough's quick thinking and blocking action, as she drove her pickup truck between Anita and the attacking moose, allowing her to jump in the passenger door. The moose had reared back and struck, like a boxer, with its front hooves, injuring Anita's shoulder.

Another friend, Dave Parker, regularly checked on the Correira family's trailer while they were out of the state for the winter, shoveling masses of snow from the flat roof to keep it from collapsing. Late one afternoon, he climbed the snow bank and put on his snowshoes, preparing to check the trailer. Hearing a loud, "Huff," behind him he didn't think twice. He dove from the top of the tall snowbank, sliding headfirst under his truck, just as the moose's hooves hit the tails of his snowshoes.

Attacks were happening all over the area as the moose, hungry and irritated, vented their frustration. One local man, named Bear, was going to check his mail. He carried a heavy caliber, over and under, double-barreled rifle; hearing a moose charging from behind, he turned and fired, point-blank, into the moose's chest, knocking the moose down on top of him. As the moose struggled to get up, directly above him, Bear shot his second bullet through the moose's head, killing it.

The classic story of the season had to be when a man went to his mailbox, again carrying a powerful weapon, and was ambushed by a moose. The moose came from the top of a high snow bank and pinned the man under him before he could react. The story goes that the man's friends, drinking coffee at the Big Su Lodge saw their friend under this moose. Each time the man tried to reach for his rifle lying nearby, the moose stomped on his hand.

They came to his rescue, but the moose would not leave. He had bagged his hunter. After a time of attempting to chase the moose off, they had to shoot it as it stood above the man on the ground.

My family and I avoided a potentially very bad moment when, late one very dark and snowy night, I drove the road near our home. I heard the word, *moose*, in the back of my mind. I quickly stopped, just as a moose came off the tall snow bank, totally unseen until he landed in the road, just in front of the car. If we had kept driving, the moose would have landed right on top of us.

WHEN NOTHING RUNS BUT THE DOGS

Propane freezes around fifty below zero. Trying to start a vehicle in that kind of cold is a near certain damaging of the engine or other moving parts. When it gets that cold, most everyone stays at home, tends the fire and tries to keep the water pipes from freezing up. Nothing runs but the dogs.

Sled dogs are powerful, happy, and fun creatures. A good one will pull and work with you hour after hour and day after day. A lazy one learns to look like they are working when doing very little.

Skiing behind sled dogs is called *ski joring*, a Scandinavian term. Often, Roni and I would travel for miles, sometimes under a full moon, Roni giggling in her joy as a dog would pull her faster and easier than she could have ever skied on her own. I would often ski with them for twenty miles or more.

They have their own language. "Hike!" is the term used for *let's go*. "Gee and Haw" for right and left turns. "On by" is used to tell them to ignore whatever distraction may be alongside the trail. Sometimes it works, sometimes it doesn't.

As I was learning the terms, I came up with a reminder for Gee and Haw. I would say to myself, *Gee, I think I'll go to the river, (a right turn) or Haw haw, I think I'll go downtown (a left turn) and let everyone laugh at me as I desperately try to control these fresh dogs.* I almost always said, "Gee."

Skiing behind more than one dog can be especially fast and thrilling. When they are fresh, these dogs run well over twenty miles an hour.

Once, I was traveling at high speed behind our two dogs over icy roads when they were suddenly distracted by a loose dog running at them from a side driveway. How they ever managed a ninety degree turn at that speed was a testimony to the traction of their toenails. They ran, at full speed, down the driveway after the dog, while I was moving at a right angle away from them, continuing to glide down the road at the same speed. We were headed for a train wreck.

Somehow, I leaned to the side and yanked on the lead rope just at the right time, yanking them, backwards, out of the driveway and back on course, without a crash.

Two dogs are usually joined at the collar to keep them flowing together in the same direction. I decided—once—to try a "Greenland hitch," or "Fan hitch," that allows the dogs greater roaming freedom by not joining them at the collar. While skiing down a hill, I outdistanced the dogs, who then came up behind me, one on each side in their new freedom and yanked me off balance as the lead rope caught my legs from behind. We crashed into the powdery snow where Sable and Silver then proceeded to jump and frolic on and over me as if they had just pulled a most wonderful practical joke!

MOOSE DROPPINGS

"No way, you're never going to get me to believe that," the newcomer to Alaska, Jack, said as we showed him the pellets that are the wintertime excrement of a moose. These strange, oblong, woody blobs about the size of the end of your thumb just don't look like a moose turd should look.

People do the strangest things with them. They paint them gold and string them together to make necklaces. Or, sell them as gold Moose Nuggets to the tourists. Or, an ultimate classic, they create Moosesquitos from the droppings and porcupine quills.

The strangest of them all has to be a moment that came about during the annual Moose Dropping Festival held in Talkeetna during the late summer. The festival, a local Chamber of Commerce event, is a time for parades, softball tournaments, kibitzing with area politicians and generally having fun. One of the popular games is the Moose Dropping game, where gold-painted moose turds are tossed onto a table with numbered squares, with the high scores winning a prize.

Gene Jenne, a proud, fun-loving Norwegian and local icon who ran the gas station, picked up the ringing telephone one day just before the festival. "What's this about a Moose Dropping Festival?" an irate voice said on the line. "Just how high do you drop these moose from?" The voice demanded, "And what are they dropped upon? I am an animal rights activist and I demand to know what you are doing to these poor moose!"

Gene, not often missing an opportunity like this one, calmly said, "We take the moose up under a helicopter and drop them from 1,000 feet onto the pavement. As the activist blustered on, Gene explained it was only a game. No real moose were ever dropped in any way. Now, thanks to the animal rights experience, you can buy a Moose Dropping Festival T-shirt showing a sad-faced moose slung underneath and waiting to be dropped from a helicopter above "Beautiful Downtown Talkeetna" at various Chamber of Commerce locations.

STUPID IS AS STUPID DOES

"It is not good to have zeal without knowledge, nor to be hasty and miss the way."

Proverbs 19:2 (NIV)

"Did you see how far he fell?" Peter Sennhauser said from the base of the rock climb, fifty feet above. Hearing his voice sounded good to me. Lying in a pile of granite boulders, I had taken a nasty ground fall while free climbing to the base of a climb called Bishop's Terrace in Yosemite National Park. How I lived through this one still mystifies me. God certainly protected this fool.

Often, travel in mountainous terrain will present unexpected challenges—it's a part of the adventure. However, to put yourself in harm's way through simple stupidity, that's another matter. This was just plain stupid.

Yosemite National Park is a rock climber's dream. The valley is world renowned for beautiful, solid granite rock, magnificent pillars and walls sweeping thousands of vertical feet above and awe inspiring views. In 1973, while still living in Alaska, I met some friends in Yosemite Valley to develop my rock climbing skills and have some fun.

Some fun! Most of it was great, but I was having particular difficulty with what is called off-width crack climbing. The vertical, off-width cracks are just what they sound like, off-width. Nothing seems to fit. No good finger holds or fist jams, only two walls of granite rock at a width that doesn't allow anything simple. It's a good place to be a contortionist, twisting and jamming various body parts into the crack and wriggling like a snake to gain an inch or two of height. I couldn't seem to manage even getting off the ground.

One morning, Peter and another friend, Robbie and I headed off to a particular off-width climb in the Bishop's Terrace area. They managed it well. I failed miserably. Frustrated and not thinking clearly, I chose to solo climb up a short fist crack to the base of the next climb. It looked easy enough, but when I reached a ledge twenty feet or so off the ground that I thought I could pull up on, I found it was sloping and sandy. Climbing back down the way I had come up was my only option.

Reversing my steps in the vertical, fist-width crack, I began to climb down, not realizing, in my inexperience, that my fist jam holds would rotate out as I moved down with weight on them. In the blink of an eye, I was off the rock in a back dive, falling headfirst into the rocks and tree roots on the ground, fifteen feet below.

I remember plainly watching the rocks and tree roots as they seemed to rush up to meet me as I fell, headfirst to the ground. Bouncing and rolling through the large granite boulders in the boulder field below, I came to a stop and, realizing I was alive, began wild and uncontrollable adrenaline laughter.

Climbing back to the base of the next climb, I thought all was well until I began to shake as the adrenaline wore off. I needed a rest. Rappelling down the rock to the ground, I lay down in a grassy meadow, enjoyed the warm sun, and watched my friends complete their climb.

It would have been nice never to be stupid again. Oh, well.

A VERY NEAR THING

Crevasses are somewhat unpredictable. Usually, they form where the stresses of gravity pull the snow and ice away from the slower moving or stable snow and ice fields behind. On high, glacial mountains, they can also form on ridge lines, when the large, heavy cornices formed by blowing snow pull away from the rock of the ridge itself. Seeing them is easy if they are not covered by snow bridges that form over the expanding crack due to wind and snowfall.

Leading an Expedition School on the Kahiltna Glacier had been fun. We had wonderful days of rock and ice climbing and had some good climbs of the smaller peaks in the region. Training for eventualities, we taught ice axe self-arrest skills, in case of a fall, and crevasse rescue techniques. Always, we stressed the need to stay roped together due to unforeseen crevasses.

Climbing Peak 11,300, unnamed at the time, but now known as Mt. Francis, situated at the confluence of the main Kahiltna Glacier and the Southeast Fork would be a graduation exercise, with the students taking the lead. Named for an eccentric and famous violin playing base camp manager, Francis Randall, Mt. Francis would offer reasonable and beautiful climbing to a scenic summit. We started off in the cold, early morning, kicking steps up a steep snow slope that led to the North Ridge.

A short, steep rock step slowed the group and they asked for help to overcome it. Unroping on the stable snowfield, I climbed around the rope teams ahead and helped them over the step. Enjoying the freedom, I climbed ahead to the ridge.

The views were expansive. Mt. Hunter, Mt. Foraker, and Denali surrounded the peak and showed themselves crystal clear snow and ice giants thousands of feet above. I was enchanted.

Now on the heavily corniced ridge, (some of the cornices on the ridges of these mountains can be 100 feet across the top), I hesitated and thought I should wait for the group below to catch up and re-rope. It was cold, they were slow, the setting was beautiful, and I got stupid.

Slowly, I picked my way up the easy ridge, noticing where another team had traveled a few days before. They walked much farther out on the cornice than I felt safe doing. Years before, I had watched from a ridge top as a massive section of cornice over one-half mile long had peeled off from the West Ridge of Mt. Hunter in an earthquake.

A few hundred feet up the ridge, I took one step on snow that felt different, more crunchy and icy. As I took the next step, my forward foot punched through the thin, icy snow that covered a hidden crevasse. I froze in my tracks. Very cautiously, not even daring to breathe, knowing my back foot was on the same, thin snow bridge, I balanced on my back foot and withdrew my front foot from the crevasse. A gust of wind from the hole my foot had made blasted ice crystals into my face as I shifted my weight and slowly stepped back to solid ground.

My heart was in my throat. I knelt there in the snow, realizing the potential consequences of my foolishness. Waiting for the students to join me, I showed them what had happened and used my stupidity as an object lesson. I hoped they would remember; I, certainly, will never forget.

Forrest Gump's mother was right; "Stupid is as stupid does."

DENALI DINOSAURS

"It's not the years in your life, but the life in your years."
--Abraham Lincoln

In 2005, it was magnificent desolation all over again—kind of. Denali's West Buttress experience had changed. It had been twenty-six years since my last climb of the mountain in 1979. After we returned from this climb, I would write to the National Park Service in Talkeetna that it was like returning to a boyhood home that had since been purchased by someone else and remodeled. Eighty people, strung along the summit ridge that late June day, made the summit a community affair. I hardly recognized it. It was still the same high, cold, beautiful mountain I had first come to know in 1974, but the social experience was vastly different.

It had all begun for us the year before, on a park bench at my mother's eighty-third birthday party. My sister Nancy asked if I felt I would ever climb Denali again. Her daughter, my niece Melissa "Missy" Lee was interested in climbing the mountain and Nancy wondered if I might go with her to lead the trip. I said I would think about it. "It will take a full year of training if we decide to do it," I warned Missy. The seed was planted. Even though it had been a long time since I had been there, the thought of returning to where I had spent much of my young life was enticing.

The last climb in 1979 had changed the course of my life. I still vividly remembered the rescues, the sickness, the storms, and exhaustion that had been overcome by the grace of God, and how my whole picture of life and its meaning had never been the same since. It was sobering to think of returning to the mountain with Missy and my two sons Jonathan and Michael along.

It was Michael's high school graduation present. Joining with my kids and me was family friend and climber, Cindy Jones, who had unfinished business with Denali, as she had reached 18,200-foot Denali Pass with a Park Service group years before, but had been turned around by a rescue situation. Finishing out our team would be Steve Nutting, climbing director for Camp Redcloud in Colorado, and his good friend, Derek Fullerton.

It was a strong team, both young and old. Cindy and I, both now over fifty years old, thought our expedition team name, Denali Dinosaurs, seemed appropriate. We were part of an earlier generation of climbers. Things had changed.

One very significant change was the requirement to poop in a specially designed can and carry it off the higher reaches of the mountain. It made all the sense in the world, as sanitation was very difficult to maintain in a region of hard ice and harsh winds. As strange as the experience was, it was far better than being sickened from tainted snow so close to the summit. On the lower glacier a plastic bag would be our outhouse. After accumulating enough waste, the bag would be thrown into a deep crevasse, keeping the glacier much cleaner and safer than in the past.

We found more changes when we landed at base camp on the Southeast Fork of the Kahiltna Glacier. The once nearly empty camp now had a full time manager and was run with the efficiency of an international airport, which in many ways, it was. Thousands of people would fly in to the glacier during the season, some to climb, others just to enjoy the scenery. Crossing the runway to work on our crevasse rescue skills nearly caused a major incident. There was a different place for that now. I experienced how different it was again as K-2 Aviation flew a thirty-member, high school jazz band to the glacier to perform for the hairy, unwashed mountaineers. No doubt, it was an interesting experience!

We began the long trek up the Kahiltna Glacier, shuttling loads of camp gear, food and fuel up the mountain in stages. Exhausted from the start, we had all burned the candle on both ends trying to get packed and onto the mountain. We passed a head cold between us on the lower mountain, giving rest days to those who were sick and causing the workload to increase for those who stayed healthy. This is never good news on a big mountain. It's difficult to recover in the cold and high altitude, and often impossible as you climbed higher, for your body just doesn't have the resources it needs to heal in the thin, cold air.

After a week of hauling loads, we were at the 14,200-foot advance base camp, waiting on weather in order to move to the high camp at 17,200 feet. Current weather was a problem, but the long-term effects of climate change were a problem, too. Rock fall at 13,200-foot Windy Corner was now a consistent and potentially deadly problem where it had never been before. The glacier had changed as well and presented some difficulties skirting some large crevasses.

We carried a load of food, fuel, and equipment and cached it high on the ridge of the West Buttress at around 16,800 feet. Then we waited—for six days. The weather was not smiling on us. We spent the time trying to hike around in the winds to get some exercise, reading books, playing cards in a deep snow cave, and enjoying an international talent show.

We figured we had a dozen nationalities, and over one hundred people, there in "Denali City." National flags flew above the camps, and exotic languages could be heard as we walked through the camp. We invited them all to an evening of impromptu talent choreographed by my niece, Missy. The Irish sang and led us all in a very strange and funny song and dance. In the sub-zero temperatures, dressed like multi-colored, fluffy chickadees in all our down gear, we sang together, "I'm singing in the rain, just singing in the rain. What a glorious feeling I'm hap, hap, happy again."

"All right, cut," the leader would direct. "Leg out, head out, bum out, tongue out... Aaroo-cha-cha, aaroo-cha-

cha, aaroo-cha-cha-cha." It was a hilarious spectacle as climbers of multiple nationalities hopped around on one leg, singing and trying to follow directions. Others told jokes; the boys from LokiGear demonstrated innovative equipment they had invented. And then there was Frisbee.

In the poor visibility of a gentle snowfall, a large group assembled to throw a Frisbee we had brought. Australians, Spanish, French, and Irish disappeared into the deep snow more than once as the result of a diving catch.

And, still we waited.

Jonathan, who had not been able to train properly and had a weak ankle from an old injury, decided to stay back and wait for us at 14,200 feet when the weather broke. He would climb to the ridge of the West Buttress at 16,000 feet later on after he rested and the weather improved.

Finally, a small window of opportunity opened with improved, but not perfect, weather. We packed up and headed up the slope above camp to the fixed lines between 15,000 and 16,000 feet. Step by step we slowly ascended the slopes below the fixed lines with our heavy packs. It was especially tiring after all that waiting and lack of exercise. Climbing to the base of the fixed lines, Cindy began a wheezing and labored breathing. As she was not recovering well, it was plain she should not continue up into the higher elevations. She would have to descend to the lower camp and wait.

We said a painful and disappointing goodbye. After all that effort, to lose a teammate this way was hard to handle, but better than the alternative of continuing up and into greater difficulty. I was reminded of what British mountaineering legend Don Whillans said to a group of young climbers high on the West Rib of the South Face many years before. Waiting out a storm for days, camped on an icy ledge on the 10,000-foot wall, the younger climbers said, "We have to move." Looking out the tent door Whillans said, "Well lads, I look at it like this. I can see fighting my way out of a bad situation, but I sure as hell can't see fighting my way into one!"

We began the ligament stretching process of moving up the steep ice, protected from a fall by clipping our rope ascenders onto the safety lines anchored into the icy slope. Crampons biting into the ice, step by step, little by little, we ascended to the ridge above carrying our heavy packs. The weather being so unsettled, we were thinking conservatively and had packed enough food and fuel to last five to six days at the 17,200-foot high camp. The long climb between camps was exhausting.

On the ridge at last, we enjoyed the expansive views of the entire Alaska Range unfolding around us. We could see hundreds of miles, over the giant peaks, to the lowlands, more than 15,000 feet below. Many other groups were taking advantage of this break in the weather, causing traffic jams at difficult sections of the ridge.

As the sun and temperature dropped, the winds picked up. We would often be stalled, waiting for other, very slow, groups to move off a slope above. One climber not doing well could stall dozens of others for hours. As the sun finally dropped behind a ridge, we found ourselves seriously tired and cold, waiting for a German woman to clear a short step of fixed rope around a rocky ledge. She was exhausted and unable to move, so we had no other option but to wait.

She secured herself to an anchor in the snow part way up. We moved through, communicating at the top of the ridge with her team, who had left her to travel alone as she was moving so slowly. They descended to help her the final quarter-mile to high camp. This was getting very awkward. It was late, the wind was up, it was cold and people were strung out all along the mile long ridge between 16,000 and 17,200 feet. I had never experienced a more likely setting for multiple disasters.

Stumbling with weariness, we arrived at high camp. Finding some old snow walls built of blocks carved from the icy snow by a previous team to protect the tents from high winds, we dug in. Setting up our dome tents, we immediately began to melt snow for water and soup as we were both dehydrated and hungry. Thankfully, it was a moderately warm storm. Temperatures were only slightly below zero.

Tired to the bone, we crawled into our sleeping bags, continuing to cook in the small vestibule area at the tent's entrance. Muscle cramps hit as our poorly fed and used-up muscles protested what we had done to them. But, we were safe and warming up.

Out on the ridge, many climbers were in trouble. One South American climber, separated from his group in the storm, stumbled into our camp in the early morning hours asking for help. "Do you know where my friends are?" he asked in weary, high-altitude confusion.

Not being able to help him find his team, I told him, "Grab your sleeping bag and jump in our tent, we can find them in the morning." Our tent was more than full, but the extra body provided more warmth, and we slept reasonably well, after tanking up on food and water.

In the morning we took stock of the situation. Everyone had made it off the ridge alive and unharmed the night before. Now, we waited once again for weather good enough for a summit bid.

For four days we waited, watching others try for the summit in the strong winds. Some climbers, desperate as they began to run out of food and fuel, took extreme risks to get to the top before they ran out of time. Bad weather reports kept coming, groups that had traveled halfway around the world for this chance quit and descended to the lower mountain and the long trip home.

On our fifth day of waiting, it wasn't looking good. We were getting dangerously low on food and fuel and the weather showed no signs of a let up. That night, the winds blew furiously as I prayed in my sleeping bag for many hours, deeply concerned for the health and safety of my crew. Cindy, feeling much better after a few days at 14,200 feet, had climbed, solo, from the lower camp to join us and help us descend. I could hear her scream in her tent through the wind noise as a leg cramp caught her. We were at the point of a critical decision. I still felt an assurance we were not only supposed to be there, but that we would stand on the summit. Finally, it came down to faith. Did I truly believe we had God's assurance we would succeed? *Yes,* I thought. I fervently and prayerfully proclaimed, *God, I've done all I can, I choose to trust you. Now, you do your work.* I then fell into a peaceful sleep, with the wind still roaring, at about 3:00 am.

At 5:00 a.m., I awoke to total silence. I poked my head out the tent door and looked around. I was greeted with dead calm and perfectly clear weather. Moved almost to tears, I shook the other tents, waking the team with, "Does anyone want to climb this thing?"

Slow getting out of camp, we watched as over eighty climbers stretched out, in one very long line, on the traverse to Denali Pass. We found a place in line and started for the summit. Cindy, who had given up her hopes to finish the climb, decided to give it a try and joined us as we cramponed up the steep slope to the pass above. It was a very long and slow-moving day. Groups not doing well with the 20,000-foot altitude would refuse to step from the trail and allow others to pass, leaving no other recourse but to wait and wait as they slowly and painfully plodded toward the summit.

Eleven hours from high camp, as clouds began to form and thunder began to sound in the lower distance, we arrived. We had made it; after all that waiting and hard work. Tears flowed from great joy and deep relief. It had taken twenty-one days to reach the summit. For me, it was completing something that had been left undone for twenty-six years. On my last professional climb with my guide service, we had been denied the summit due to rescues and storms. Now, we enjoyed the summit view in an unheard of and balmy fifteen degrees, with dozens of others, a very different experience for me. Over eighty climbers, stacked up waiting for the weather, reached the summit that day.

Summit photos taken and congratulations complete, we began the long descent. The weather, warmed as it was from thunderstorms developing below us, closed in. We kept moving, dropping down through a disorienting and dizzying combination of blowing snow and cloud. Finally, after seventeen hours of climbing, we returned to our camp as the thunderstorm weather began to clear. At the foot of the mountain, in Talkeetna, the major thunderstorm had caused considerable tree damage. Winds toppled trees and hailstones stripped leaves. We had been descending through the upper portion of a large thunderhead.

Hearing a weather report of another storm moving in, we were up early and climbing down the rocky ridge to the fixed lines and Jonathan's camp below. Jonathan, who had been alone for a few days now, met us with hot grilled cheese bagels. We ate, drank, and rested there before a full night's descent to base camp. Near the summer solstice, in late June, the sun was still brightly shining on the northern horizon at midnight.

Shuffling the long miles down the glacier on our snowshoes and pulling our sleds behind us, we raced the weather to base camp. It was still beautiful as we pulled into camp at ten in the morning. We had been traveling for fifty-two hours with only a few hours rest. Climbing the last miles of Heartbreak Hill, the uphill slope of the Southeast Fork of the Kahiltna Glacier, to the landing strip, I had fallen asleep while walking, waking up three different times with a sudden start, thinking to myself, *Man, I hope I'm walking in the right direction!*

We had buried a cache of food three weeks before at base camp. In that emergency stash we had left fresh fruit: pears, apples, and oranges. After twenty-two days buried in the snow, they were as fresh as when we bought them and more than delicious.

As we flew out, the storm developed, and we were able to spend a few days in Talkeetna, eating ice cream, moose burgers, and pesto sandwiches, and sleeping in a wonderfully dry and warm cabin instead of spending more stormy days on the mountain.

MORE BEYOND

"Ah, but a man's reach should exceed his grasp, Or what's a heaven for?"

--Browning

In Portugal there is a monument that depicts a lion scratching out the word "No," from a sentence that says "No More Beyond." It is a monument to that city's native son, Christopher Columbus. Prior to Columbus's voyage to the New World, it is said there were pillars placed at Gibraltar, the passage out of the Mediterranean Sea into the Atlantic, that warned mariners to be prepared as, it was thought, there was no more land beyond that point. Columbus discovered that there was "More Beyond" and the world's thinking was changed forever. So it has been for me.

Where at one time I fell in love with the awesome beauty and power of the high mountains of Alaska and the world, I now knew that there was "More Beyond." My experiences with the Creator of the mountains would forever change me and how I viewed the world around me. While I still greatly enjoyed them, the mountains themselves were no longer my source of peace and purpose, but the one who made them was. He could be found anywhere, at anytime, if I was living in awareness of his presence.

These experiences, and the lessons learned through them, have carried me through the many joys and challenges that life offers, the births and deaths, the failures and successes, and the comings and goings of careers and relationships. I have found the author of life, Jesus Christ, to be as sufficient in the challenges of parenthood or with the tribes of the Amazon Basin as he has been for me on the slopes of Denali.

The Tanaina Indians of Alaska called the mountain "Denali, the Great One." I found the Lord of the mountain to truly be "the Great One," and I will be forever grateful for the wonderful experiences, friends, and grace given to me through the high mountains and wilderness of Alaska.

> "I will lift up my eyes to the hills, where does my help come from? My help comes from the Lord, the Maker of heaven and earth."
>
> Psalm 121: 1-2 (NIV)

> "Your love, O Lord, reaches to the heavens, your faithfulness to the skies. Your righteousness is like the mighty mountains, your justice like the great deep."
>
> Psalm 36:5-6, (NIV)

EPILOGUE: IT'S AN AMAZING MYSTERY

The telephone call came on Sunday afternoon. I was volunteering as a chaplain at a local hospital and a family was requesting my presence as they removed an older family member from life-support. Not knowing what to expect, but anticipating a comatose patient whose body systems had shut down, I drove to the hospital.

What I found as I arrived at intensive care was anything but comatose. I walked into a room of swirling emotions. The patient, Stan, (not his real name) an eighty-year-old man, was totally awake, anxious, and aware of the moment. His children were sorrowful but stoic, his grandchildren distraught and weeping. The situation was overwhelming.

I was told by the family that Stan's organs had shut down beyond recovery and that they would soon be turning off the many machines he was attached to. The family was distraught and overrun with pain and confusion.

Stan had been a hard-working farmer his whole life. Work was his life. Since the Great Depression there had always been more work to do than there was time to do it. He had never spoken to his children or grandchildren about his faith or anything spiritual. Life was just about work. The family loved their Dad and Grandpa but had no spiritual or emotional tools to help them through this moment.

He couldn't speak, being on a breather. I asked the family for permission to speak to him. After introducing myself I said to him, "Stan, you know what is going to happen here. These machines are going to be turned off and your body is either going to work or it's not. Are you ready for that?" With downturned face and eyes he shook his head to say, "No." It was plain he was deeply troubled.

I could see he had a need of forgiveness and asked if that was so. He nodded, "Yes." Explaining the love of God to him, I told him of the mercy and forgiveness offered to us through God's Son, Jesus Christ. And, that his death on the cross on our behalf covered all of our failures. When I asked if he would like to receive this forgiveness and cleansing, he nodded enthusiastically. We prayed together, and as we finished, Stan looked up, straight into my eyes, and I could see faith, hope, and love had replaced fear, guilt, and shame.

He was alive in a whole new way.

We all read Psalm 23 together, "The Lord is my Shepherd, and even though I walk through the valley of the shadow of death, I will not fear, you are with me," and left the room as the nurses came and removed the life-support. As we waited in a room off the hall of the ICU, the nurses prepared us for the coming minutes and what to expect.

As we returned to Stan's room we immediately noticed he was not going downhill, but getting stronger minute by minute. Within two hours he was sitting up, eating cheesecake, and drinking a soda. When I came back the next morning I found him joking and laughing with his respiratory therapist. He looked at me with eyes full of amazed joy and said, "Shouldn't I have a Bible?"

His vital signs and the condition of his internal organs had not changed. He shouldn't be alive. Not knowing what else to do, the hospital allowed him to stay in a regular room for a week while he visited with and loved on his family, read his Bible and ate butterscotch ice cream. The nurses kept telling me, "We don't understand, according to our tests, this shouldn't be happening." After those special days, Stan slipped away in his sleep without trauma.

This life, and the life beyond the natural is a great mystery, but it's real. Hang gliding, rock climbing, scuba diving, and high-altitude mountaineering are all wonderful and exciting adventures, but to me they cannot compare with the adventure of following a living God day-by-day. Faith is the highest adventure of all.

Life is an adventure. Impossible situations come upon us more often than we would like. Finding God in those moments is as simple as a prayer for help. He is always faithful and always available.

Here's to your *Highest Adventure.*

--Jim Hale

GLOSSARY OF TERMS

Ascenders: A mechanical friction device used to ascend ropes.

Col: A large, flat saddle on a ridge.

Cornice: An overhanging deposit of wind-transported snow that forms on the leeward side of ridges.

Crampons: A frame of metal spikes attached to a climber's boot to provide increased traction on snow and ice.

Crevasse: A narrow and deep crack that develops in the surface of a moving glacier as the glacier moves over uneven terrain. Many are concealed by covers of snow called *snow bridges*.

Fixed Rope: Ropes secured into rock or ice semi-permanently to provide security to climbers in case of a fall.

Glissade: To slide down snow slopes in a controlled way using an ice-axe to brake.

Ice Axe: A climbing tool with a spike on one end and an adze head and pick on the other to provide a hold on ice, chop steps, or to perform self-arrest.

Ice Hammer: A short hammer with a pick on one end used to provide an anchor on steep ice.

Ice Screw: Threaded aluminum tubes used to provide anchors in ice to secure a climber in case of a fall.

Kayak: Small one- or two-person covered boats.

Moraine: Bands of rock and gravel transported and deposited by the movement of a glacier. Lateral Moraine on the sides, Medial Moraine in the middle, Terminal Moraine at the snout, or end of the glacier.

Rappel: To descend by sliding down a rope, usually done with a mechanical friction braking device.

Self-Arrest: To catch yourself in case of a fall using the pick of the ice-axe to stop you or catch a roped-up teammate.

Snowshoes: Frames of webbing strapped to boots to distribute weight in travel over snow.

Wind Blast: The force of wind built through the momentum of an avalanche. Also known as a powder blast.

ABOUT THE AUTHOR

Jim Hale was born in the Territory of Alaska in 1953. One of the early mountain guides on North America's highest mountain, Mt. McKinley (Denali), Jim is the Senior Pastor of Spirit of Life Christian Fellowship in Grand Junction, Colorado. He lives with his wife of thirty-five years, Roni, in Palisade, Colorado, where they continue to adventure with their three grown children and two grandchildren.

Jim and Roni are available to speak at men's meetings, churches, and schools. They may be contacted through their websites, jimhale@tateauthor.com and highadventureliving.com